Manufacturing Technology and Production Management

Manufacturing Technology and Production Management

Edited by Jeff Hansen

CLANRYE
INTERNATIONAL
www.clanryeinternational.com

Clanrye International,
750 Third Avenue, 9th Floor,
New York, NY 10017, USA

ISBN:978-1-63240-614-9

Cataloging-in-Publication Data

Manufacturing technology and production management / edited by Jeff Hansen.
p. cm.
Includes bibliographical references and index.
ISBN 978-1-63240-614-9
1. Production engineering. 2. Manufacturing processes. 3. Production management. 4. Business logistics.
5. Manufacturing Industries--Technological innovations. I. Hansen, Jeff.
TS183 .M36 2017
670--dc23

For information on all Clanrye International publications
visit our website at www.clanryeinternational.com

Printed in the United States of America.

Contents

Preface

The purpose of the book is to provide a glimpse into the dynamics and to present opinions and studies of some of the scientists engaged in the development of new ideas in the field from very different standpoints. This book will prove useful to students and researchers owing to its high content quality.

This book on manufacturing technology and production management deals with the processes that determine how products are to be manufactured. Various kinds of production processes exist which manufacture particular commodities; some may rely more on technology and others on personnel. Time and cost estimates play an important role in setting goals and strategies for the same. This book includes contributions of experts and scientists which will provide innovative insights into the field of manufacturing and production management. Contents included in this text aim to contribute to the already existing research on production management and technology. It traces the progress of this field and highlights some of its key concepts and applications. This book, with its detailed analyses and data, will prove immensely beneficial to professionals and students involved in this area at various levels.

At the end, I would like to appreciate all the efforts made by the authors in completing their chapters professionally. I express my deepest gratitude to all of them for contributing to this book by sharing their valuable works. A special thanks to my family and friends for their constant support in this journey.

Editor

Reducing Internal Information Technology Resource Allocation Through Global Upstream Electronic Business Standards

Douglas Steven Hill
University of Southampton
dsh1e08@soton.ac.uk

Juan Francisco Zurita Duque
Novozymes A/S
jfjd@novozymes.com

Helle Skøtt
Novozymes A/S
hstt@novozymes.com

ABSTRACT: An increasing number of organisations are beginning to operate in a standardised business process environment suggesting that there is benefit in a standardised business messaging infrastructure.

Novozymes A/S is the first known industrial biotech organisation in Denmark to apply the GS1 and the Consumer Goods Forum's global upstream standards initiative (GUSI) in a Vendor Managed Inventory (VMI) scenario. The combination of applying a standardised VMI business process using the aforementioned integration standards for electronic business messaging and pre-agreed standardised message choreography has been proven by Novozymes to add significant business benefit to their organisation.

This case study methodology outlines the results of Novozymes' application of GUSI standards over three separate implementations replenishing 25 customers' factories and describes the resultant reduction in internal IT effort. The implementation data were collected from a total of twenty five factories (sites) which Novozymes, today, replenish using GUSI VMI. The 'ex post' results were then compared and interpreted against a Novozymes internal benchmarking analysis which was used an 'ex ante' base line.

The findings strongly suggest that the GUSI VMI application provides not only reduced integration effort, but also is a foundational basis for higher rent generating processes and improved demand transparency management.

Keywords: GUSI, Supply chain, Integration, Visibility, Standards, VMI

1. INTRODUCTION

1.1 Competing Supply Chains

Global eBusiness standards are required by manufacturers and their suppliers to improve transparency in supply chains to enable higher rent generating processes (Angeles, Corritore, Basu, & Nath, 2001; Fabbe-Costes, Jahre, & Rouquet, 2006; D S Hill, 2009; Hau L. Lee, V. Padmanabhan, & Seungjin Whang, 1997; Lee & Lim, 2005; M E Porter, 2001; Rodon, Ramis-Pujol, & Christiaanse, 2007; Smith, 2003).

However, as much as the application of a standard process or a standard electronic messaging guideline in a dyadic trading relationship is useful, standards are more constructive when rolled out across the value chain where economies of scale can be leveraged and internal effort optimised (Hsieh & Lin, 2004. p. 71). Indeed, with the emergence and subsequent consolidation of trade globalisation (den Butter & Linse, 2008; Hertz & Hultman, 2008; IBM, 2009; Iskanius & Kilpala, 2006), companies compete not against each other, but against supply chains on a international scale (Christopher, 1992; Lambert, Cooper, & Pagh, 1998) .

This trend would suggest that relationships between value chain stakeholders should aim to collaborate in order to reduce total cost within their trading environment. Value chains can be optimised for effectiveness by the implementation of certain vertical integration scenarios such as Traditional Order Management (TOM), or enhanced using processes such as the Collaborative Planning, Forecasting & Replenishment (CPFR®) framework (Akkermans, Bogerd, & van Doremalen, 2004; Barratt & Oliveira, 2001; Skjoett-Larsen, Thernøe, & Andresen, 2003; Småros, 2003; VICS, 2010). TOM, however, can be improved upon through a more intricate collaborative scenario such as Vendor Managed Inventory (VMI) which is designed to smooth the flow of goods and reduce the bullwhip effect ultimately enhancing the economies of involved parties (Disney & Towill, 2003; Holmström, 1998; H. L Lee, V Padmanabhan, & S Whang, 1997; Hau L. Lee, et al., 1997; Lehtonen, Småros, & Holström, 2005).

Nevertheless, where trading party collaboration accords the opportunity to improve supply chain activities, Porter's five forces model of competition observes that one trading partner will, more than likely, hold the power in a trading partner relationship (2008). This would suggest tension between the actors and introduces the perennial themes of trust and risk which need to be addressed before successful business integration can be achieved (Lee & Lim, 2005).

This type of tension can, under some circumstances, be mitigated, reduced, or turned into competitive advantage through the use of standards organisations which can act as neutral brokers in the development and maintenance of supply chain standards and processes. Porter goes as far to say that a great deal of the economic value created by marketplaces can be traced back to the application of standards (2001).

This case study reviews the application of the VMI scenario implemented by a global enzyme manufacturer based in Denmark, Novozymes A/S. It details the benefits of harmonised IT integration and discusses the global upstream standards they have applied in several similar implementations.

Specifically in focus in this review are the internal technical resource allocation and benefits realised through GS1's XML (extensible mark-up language) based electronic data interchange (EDI) standards and aligning message choreography. These elements, which have been experienced by Novozymes, facilitate the improvement of business processes and reduce integration costs between participating value chain members (D S Hill, 2009).

2. LITERATURE REVIEW

2.1 Global Backdrop to Trade Pattern

2.1.1 Globalisation: Patterns of global supply

The growth of globalisation, as emphasized by Halldorsson *et al.*, is illustrated through a report which states that $24 trillion US dollars worth of imports and exports were recorded by the WTO in 2006, nearly double the figure reported in 2001 (2008). Specific emergent patterns of global supply can now be detected (BERR, 2009; EC, 2010a; A Halldorsson, et al., 2008; IBM, 2009) which are materialised as organisations take advantage of low cost manufacturing, or production facilities, across the world. Western organisations are increasingly taking note of emerging markets in India and China and are aggressively expanding their presence in these regions (Pankaj Ghemawat & Hout, 2008). Lessard, however, warns that other contributory factors are at play when organisations choose locations for outsourcing production and that best value should be the criteria for choice rather than purely low cost (2008). Other research takes a contrary view to the popular belief that humankind is in a globalised community. Ghemawat proposes that commerce is actually in a state of semi-globalisation and, in essence, borders still matter and that internationalization is less advanced that we may think (2006).

Nevertheless, whichever view is ultimately correct, globalisation and the increase in cross-border trade is now a growing trend and is viewed as a path towards competitive advantage through global sourcing (den Butter & Linse, 2008; Hertz & Hultman, 2008; IBM, 2009; Iskanius & Kilpala, 2006). Globalisation therefore is a real enough phenomenon where extended supply chains are increasingly affected by a growing number of logistics actors making tracking and tracing products over longer distances more difficult. Butner notes that the number of transnational organisations in the world doubled between 1995-2007 and the number of subsidiaries tripled, thus suggesting that the effect of increased globalisation is making the ultimate supply chain ever more complex (2010).

So then it follows, globalisation develops the concept of trading bloc economic power, stakeholders ultimately reliant on the whole bloc being as effective and efficient as possible. The North American Free Trade Association (NAFTA), the European Union (EU) and as of January 2010, the Association of South East Asian Nations (ASEAN) are the current economic powerhouses. Are they rivals? Of course, but also extremely inter-dependent as shown by China's exports in 2007, seventy percent of which were bound for Europe (WTO, 2008).

Trading blocs have recognised this interdependency and have begun to put in place initiatives to strengthen their regions. The European Commission's i2010 initiative (Europe's Information Society, 2010) is aimed at making Europe a robust trading bloc through the use of modern technology and, to this point, global supply standards are beginning to play an ever more important part of management strategy. In this macro competitive environment, trading blocs are competing against each other for business and so, to be competitive, organisations have to be interdependent within their supply chain to reap the benefits of economies of scale offered through shared standards and infrastructure.

This issue is reflected in the following pages through the lens of the Global Upstream Supply Initiative (GCI, 2010), as developed by a joint initiative from Con-

sumer Goods Forum (formerly known as the Global Commerce Initiative) and GS1. This paper reviews the benefits of applying standards in an interdependent supply chain through the perspective of a large Danish industrial biotech company, Novozymes A/S.

Novozymes have aligned many of their foundational processes and harmonised some of their business process standards through collaboration and commonly shared infrastructures. These collaborations are realised through the brokerage of standards bodies such as GS1 and the CGF which both support Novozymes by developing common GS1 XML EDI formats (D S Hill, 2009) and business information models, such as the Global Upstream Information Model (GCI, 2010).

2.2 The Global Upstream Supply Initiative - GUSI

The aim of the Upstream Integration Model and the related messaging standards is to provide tighter integration of supply chains without the need for costly and time-consuming customised IT integration projects with every partner. (GCI, 2010).

GUSI was launched in 2004 under the banner of the Global Commerce Initiative (GCI) by a group of companies in Consumer Packaging Goods (CPG). They were interested in investing in the creation of a new business and technical standard targeted on the upstream sector of the supply chain. Since then, The Consumer Goods Forum (CGF) was created in 2009 out of a merger between CIES - The Food Business Forum, the Global Commerce Initiative (GCI) and the Global CEO Forum, which, in total, has impressive combined sales of 2.1 trillion Euros (CGF, 2010).

The CGF, together with the GUSI Working Group and GS1, have successfully designed an upstream process model and the supporting XML business message standards for manufacturers of consumer product goods and suppliers of packaging, ingredients and raw materials.

2.2.1 Key Concepts of GUSI

Upstream integration is all about improving operations by sharing information and improving visibility of demand through the harmonisation of processes and standards. To this end, GUSI standards support the most common business processes used by upstream suppliers and their trading partners. GUSI consists of harmonised application of processes, GS1 product and location keys and GS1 XML business messages. These standards are a main reason why systems and processes driven by GUSI are optimised and are able to be implemented in a cost effective (profitable) and timely manner (GCI, 2007).

2.2.2 The alignment trap

Improving profitability within a value chain is a fundamental element of business (Porter, 2008) in an environment where globalisation is a growing factor for consideration in supply chain management. Globalisation is putting pressure on extended and complex supply chains making the need for global standards an ever more pressing issue where patterns of global supply are changing (BERR, 2009; EC, 2010b; IBM, 2009). Whilst there is a natural inclination for sales departments to want to comply to the buyers tender requirements, which usually means adopting their standards for the relationship, the pressure to adopt differing EDI standards and customise processes can lead to actual increased costs through a process termed the 'alignment trap' (Shpilberg, Berez, Puryear, & Shah, 2007).

Porter's (2001) and Hill's (2009; 2008) work indicate that using [foundational] standards to create integrated systems that are specifically customised for an organisation *can* enable a competitive advantage, however, whilst adapting in-house processes or other trading partners' processes and standards, Shpilberg *et al.* maintain that aligning poorly performing IT infrastructures to a business objective will not get the objective accomplished (2007).

Shpilberg's observation is instructive as it reflects the reality of business as it is today in many cases. By constant alignment with buyers' requests, inconsistent systems become complex and lose the reusability aspects of standardised solutions, economies of scale are lost to customised, individual requirements (2007).

CGF representatives from Japan reported that one large Japanese company had in excess of one thousand electronic order profiles for around a thousand trading partners, and, this inefficiency was becoming a problem! (Shibata, Ariga, Suga, & Sato, 2007). This issue is the essence what Shpilberg calls the 'alignment trap'(2007), where organisations cater to trading partner requests for IT alignment. Shpilberg notes that studies show that by aligning IT in a non-structured manner has led to organisations under-performing vis-à-vis those organisations which have applied a standardised approach by as much as 34% increased compounded annual growth over three

years (2007). In an attempt to latch onto standards' alignment efficiencies and economies of scale, Novozymes opted to apply GUSI VMI.

2.3 Novozymes Application of GUSI

Novozymes is a global organisation and appreciates that supply chains are becoming increasingly international in character. Participating in a global and complex business environment was proving to be growing constraint on their inter-company operational efficiency.

Whilst endeavouring to minimise the '*IT alignment trap effect*' (Shpilberg, et al., 2007), Novozymes and three downstream manufacturing trading partners implemented the GUSI standards. The application of GUSI necessitates the need for greater inter-organisational relationships with trading partners and standards organisations. This is something that Novozymes has embraced and as a result, has realised supply chain improvements in several areas Table 4-1.

One such benefit can be seen in reduction of internal IT connectivity effort through standardised processes where each successive implementation should become easier for Novozymes through their familiarity with the standard and of course, the back office interface is already in place and embed to VMI routines.

2.4 Standards & Supporting Role Organisations

The adoption of standards in eBusiness is primarily to ease implementation and to share development costs, however, gaining consensus or standardising, is not, as a rule, an easy task. The complexity of developing a solution generally increases through any rise in the number of stakeholders which are part of the standardisation process. This extended complexity is offset against the increased value of the final solution through a greater number of users and the final process simplification. Essentially this is the trade off when developing and applying eBusiness messaging or process standards in larger groups, although as Leonardo da Vinci is said to have put it, "*Simplification is the ultimate sophistication!*"

The neutral broker and standard governance body is then the role that standards organisations play in combination with the stakeholders that apply the standards. The aim is to reduce development costs, reduce the risk of non-adoption by trade and essentially achieve more working together, enable more advantages than through their own efforts (A Halldorsson, Kotzab, Mikkola, & Skjøtt-Larsen, 2007. p. 287).

Figure 1: GUSI perspective

A GUSI perspective in both TOM and VMI scenarios

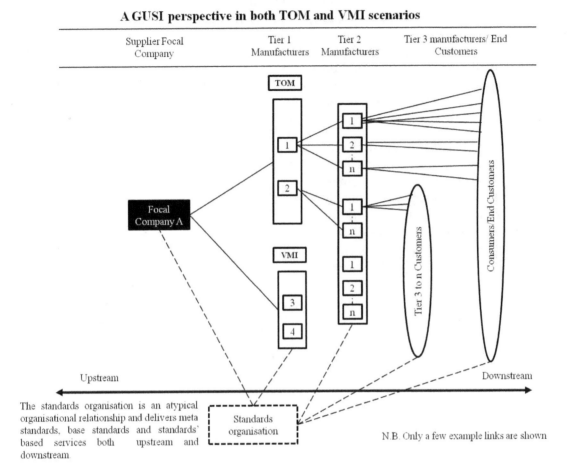

Adapted and customised by the author from: (Lambert, et al., 1998)

The role that standards organisations play in the supply chain is classified by Lambert as a 'supporting member' (Lambert, et al., 1998). These are organisations such as GS1, United Nations Centre for Trade Facilitation and Electronic Business (UN/CEFACT), International Organization for Standardization (ISO), European Committee for Standardization (CEN) and the CGF run parallel to the entire supply chain and serve the entire trading partner community irrespective of the higher level processes, such as VMI, which are in place (see Figure 2-1). A major component of some standards organisations, e.g. GS1, UN/CEFACT and CEN, as supporting members of the supply chain, is the development of electronic business message models, syntax profiles and process descriptions. GUSI standards, whilst being generally available to members as a guideline, still have to be understood by the adopter and the EDI interface to the back office applications needs to be written. However, once 'learned' and the interface

written, this expertise could be described as a meta-competency (Liedka, 1999) that adds to the portfolio of a company's resource expertise .

The value of this case study is to show that collaboratively developed standards, when in this case applied in a vendor managed scenario, equate to improved financial, services, operations and administration processes. It not only defines best practice, but also presents implementing organisations the opportunity to create a competitive advantage by reducing the alignment trap effect (Shpilberg, et al., 2007) and also as a spring board to higher level capabilities through standards and standards bodies.

2.4.1 The role of the standards & the standard's organisation – A neutral broker

Standards *development* and *use* cannot be separated (Fabbe-Costes, et al., 2006). The development of the standards should be made with consent and active participation of the users involved and the collaboration between two supporting organisations who are promoting the GUSI initiative, GS1 and the CGF, was born from the need to do just this. One of GS1's primary directives in their Global Standards Management Process is that all development should be user driven (GS1 in Europe, 2010) which mirrors the development edict from the CGF, creator of the GUSI model (GCI, 2008) .

Figure 2-1 highlights the role that standards and other supporting members play in involving supply chain members in a downstream (in Novozymes' case) TOM or VMI scenario. The key aspect here is that a focal company cannot normally influence or control the standards or processes of all links in their chain. There are suppliers' suppliers, or customers' customers for example where 'managed [eBusiness standards] links' are not appropriate or not possible (Lambert, et al., 1998). This is where standard organisations and other supporting bodies like CGF can play a role as neutral brokers in eBusiness standards development and as an implementation facilitation body. These standards organisations, such as GS1, are not necessarily bound to any specific sector vertical and can facilitate horizontal standards adoption.

3. METHODOLOGY

3.1 Study background

The research takes the shape of a longitudinal study initiated in 2003, when Novozymes started their GUSI VMI implementations. Whilst the data collected were continuously being used to optimise Novozymes' supply chain operations, this particular data set, as seen in the benefits matrix Table 4-1, was collated in 2008 for an initial draft of a forerunner to this paper. Subsequently the data have been refined and used as the main input for this research.

The data were collected from their large customers which came from the Consumer Packaged Goods (CPG) industry within House Hold Care and includes twenty five GUSI VMI supplied factories and is measured centrally based on data from Novozymes' central SAP system.

3.1.1 Study method

Novozymes made the assumption that when applying GUSI and VMI, the commercial benefits for Novozymes implementation of vendor managed inventories had been analyzed and achieved. The

supply chain, logistics benefits or Master data alignment were not considered when analyzing the IT effort business case.

The method of data collection was a collaborative effort between the first, second and third authors, the latter two are involved in applying the standards and scenarios at Novozymes. The data collection took the form of unstructured interviews with Novozymes GUSI team lead for SAP solutions and integration and the Novozymes integration manager (NB the data stems originally from Novozymes' central SAP system).

3.1.2 Entity of analysis

The study is conducted at the level of the firm and, as a result, the entity of analysis is the focal company's dyadic trading relationships with manufacturers that have implemented the GUSI process in a VMI scenario. The benefits under study are those relating to those associated purely with Novozymes' own internal operations, specifically in the area of the technical IT effort expended in establishing EDI (business messaging) connections and VMI.

3.1.3 Level of analysis

Whilst the level of analysis is primarily at the operational level, there are definite strategic aspects to the paper as a whole. Whereas the in-house departmental resource allocated to the job of IT set-up between trading partners deals with operational elements, the conclusions of the case study discuss a more strategic perspective delivered by the application of standard processes and EDI elements in the GUSI model. The paper expands the discussion of basic standards' benefits beyond simply reporting the results of the implementation. It offers a rationale as to why GUSI adoption improves strategic options, essentially using GUSI standards and standards organisations as a springboard for higher rent generating processes.

3.2 Study Limitations

3.2.1 Benchmarking

The case's results are not benchmarked against other organisations so a comparison to the actual value of the gains/benefits reported cannot be applied to an industry sector index. The study is also a cross sectional study which gives the reader an indication of the benefits in a given time frame. However, as the

GUSI model is increasingly adopted, a virtuous circle of standard harmonisation could be expected where the subsequent integration efforts with organisations already applying GUSI with another supplier should in fact speed up implementation even more.

3.2.2 Internal IT effort

No attempt is made to include or discuss benefits, or indeed disadvantages, that could affect Novozymes or their trading partners outside of the scope of the case study. The study is conducted using Novozymes as the focal company as the entity of analysis. Specifically, the case describes the amount of internal IT effort required to create electronic integration between Novozymes and three trading partners and reflects on the resultant improvement in Novozymes' performance.

4. RESULTS

The results of the case study are synthesised in the following benefits matrix which stems from Novozymes' *ex-ante* internal benchmarking against *ex-post* GUSI and VMI implementations from three strategic customers, a total of twenty five factories (sites) which Novozymes, today, replenish using GUSI VMI.

4.1 Novozymes GUSI with VMI Benefits Matrix

Table 1: Benefits matrix

Service		
Product availability	Fewer stock outs at customer manufacturing plants	1
Trading partner relationship	Improved customer relationship	3
Operations		
Truck fill rate	Optimisation through VMI management	2
Production planning	Fewer changes to the production plan	1
Trading partner integration set up	Quicker on-boarding of new trading partners and related cost savings	3
Administration		
Forecasting accuracy	Reduction in the "bullwhip effect"	2
Reduced rush orders	Increased visibility into future demand reduced costs related to order management	1
More efficient transactional message processing	Stability in, and timing of, message processing	2
Financial		
Manage working capital	Improved cash flow (less finance tied up in stock)	2
Holding costs	Reduction in inventory holding costs and warehouse management	2
Capacity costs	Increased levelling, improved capacity utilisation	1

(Duque & Skøtt, 2009)

The results in Table 4-1 are based on both tangible benefits (e.g. number of rush orders, forecast accuracy and truck fill rates) as well as the intangible benefit from 'improved customer relationship'. Whereas tangible benefits are measured over time, starting three months after 'go live' with GUSI VMI at a new customer site, 'improved customer relationship' is based on quotes from customers, the development in revenue, customer satisfaction surveys as well as supplier awards received from being a highly rated supplier of supply chain services.

4.2 Area of Benefit

The areas of benefits in Table 4-1 have been graded into three classes ranging from 1 – 3; one shows some benefit, two good benefits and three excellent benefit improvements in the different functional areas.

The discussion in the following section will only focus on the benefits of the *'Trading partner integration set up'* which is an area where 'excellent' improvements were reported in Novozymes' area of Operations.

Figure 2: Technical IT effort, with and without GUSI

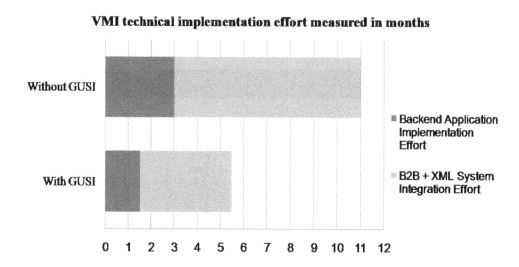

Trading partner integration set up improvements (Duque & Skøtt, 2009)

Figure 4-1 shows the implementation effort (in months) using GUSI versus a non-GUSI implementation. The integration and IT effort cover two main areas, namely the Backend application and the B2B integration and are described as follows:

Backend Application: Is the system landscape over which the supply chain optimisation processes was executed. This involved the three following systems in Novozymes' analysis:

– Enterprise Resource Planning (ERP): SAP R/3 & SAP ERP Enterprise Central Component (ECC) 6.0

– Business Warehouse and Business Intelligence (BW/BI): SAP BW & SAP BI

– Supply Logistics Planning and execution: SAP APO (Advanced Planning Organisation)

B2B Integration System: Is the technical landscape

executing the process integration management and associated services. For example: System connectivity, message transport, message routing, message and data conversion, data mapping and test monitoring activities. The Integration system involved in the Novozymes scenario:

– B2B Connectivity Landscape: SAP Business Connector, SAP Web Dispatcher and third party AS2 adapter.

– Integration Engine: SAP XI/PI (SAP Exchange Infrastructure/Process Integration).

4.2.1 Results of Trading Partner Integration Setup

Above and beyond the pure time saving aspects to the GUSI VMI implementation, Novozymes realised three major improvements as a result of trading partner integration.

1. **Process alignment:** Partnership agreements and line of business adoption have been shown in Novozymes' GUSI roll-outs to take less time than non-GUSI roll-outs. Whilst the initial start up of GUSI VMI is more complex than in a TOM scenario, GUSI partners now have compatible business processes and terms management making the management of the whole supply process simpler and less complex to manage.

2. **Promoting the GUSI concept to trading partners:** GUSI provides an industry recognized framework for trading partners that are newcomers to the initiative. Novozymes found it easier to 'promote and communicate' the GUSI concept to trading partners because the standardised framework is scalable within the industry and based not on a proprietary Novozymes solution.

3. **Technical connectivity:** Some benefits have been realised by utilizing service providers that have adopted the technical GUSI solution within their service portfolio and have now developed standardised, 'off the shelf' solutions. These solutions have saved both time of development and have been applied as scalable elements across new implementations.

The results in Table 4-1 and Figure 4-1 are the consequence of the internal calculations from Novozymes'

team lead for SAP solutions and integration when applying GS1 GUSI in a VMI environment. They are based on *departmental* timeliness and internal efficiency in establishing connectivity as opposed to an *individual FTE* cost reduction analysis. Ultimately, at a pragmatic level, a reduction in integration time must also strongly point towards a reduction in transaction costs economics and associated benefits that go along with quicker on-boarding of customers.

From the Novozymes implementation effort graph, Figure 4-1 it can be seen that there are substantial internal time saving benefits to be gained from implementing GUSI as opposed to customising solutions in a non-GUSI implementation. These benefits can be expected to increase with the number of connections and reduce transaction cost associated with trading partner integration. Figure 4-2, part of a CGF's findings, illustrates the reduction of monetary investment associated with transaction costs when GUSI is applied.

The CGF's 2006 report (GCI, 2006) findings corroborate, to some extent, the benefits indicated by Novozymes' results, albeit with the caveat that Novozymes have not explicitly indicated in Table 4-1 *ex-ante* non-GUSI implementation effort.

Figure 2: Total investment costs as a function of number of connections

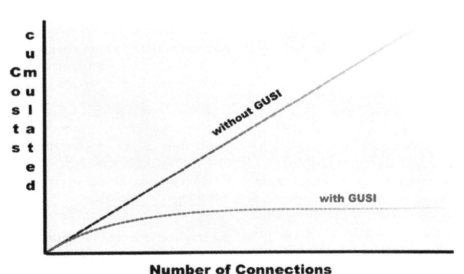

Source: (GCI, 2006)

The reduction in IT effort when using GUSI, as opposed to a similar implementation not using the GUSI standards, were calculated by Novozymes at a 1:3 ratio in favour of the GUSI application Figure 4-1. This ratio is also mirrored in an independent analysis with Carlsberg A/S who are also in the process of a GUSI roll-out (D. S Hill & Oscarsson, 2010).

5. CONCLUSION

5.1 Operations benefits

5.1.1 Internal IT effort reduction

The GUSI information model is closely tied to electronic data interchange (EDI) in the form of the GS1 XML business messages that deliver the information payload. Generic benefits of EDI are already well documented in extant literature and are not addressed in this Novozymes implementation results and benefits: See (Angeles, et al., 2001; D S Hill, 2009; Millen & Ukena, 1995; Michael E. Porter, 2001; Rodon, et al., 2007; Smith, 2003).

5.1.2 VMI Replenishment Performance Benefits

In Novozymes' experience, to successfully adopt the concept of VMI companies must consider both the technical aspects as well as business processes. VMI can be implemented as a manual process, but without the support from a technical setup it will not be possible to gain the transactional and resource based benefits (Figure 4-1). GUSI VMI considers both and additionally, it standardises business processes as well as messaging choreography. Without these prerequisites, Novozymes feels VMI would not be scalable and the true benefits from an optimised supply chain may not be achieved.

For VMI to work efficiently, a close partnership was required between Novozymes and their customers. As stock availability information is now shared and made more readily available, uncertainty and risk in the supply chain has been reduced and has led to a minimization in unforeseen events, such as demand variability. Additionally, the improved information flow has allowed Novozymes the flexibility to better plan deliveries to customers and enables the optimisation of transportation.

5.1.3 Reflections of VMI adoption

Depending on the rules and terms defined for the VMI partnership in the agreement, benefits from

increased efficiency are experienced from both the customer and supplier perspective. In environments with stable off-takes, high forecast accuracy and repetitive, high volume products, such benefits are highly visible. In these cases, the purchasing/planning function on the customer side will have no, or only little, hands on daily operation – in VMI the vendor takes on the responsibility and effort of ensuring that stock is available at the customer's site.

5.2 Standards Releasing Benefits

Whilst organisations seek divergent operations and strategy from their sector norm, the case study points towards standards as being fundamental in achieving internal operational efficiencies as well as having strategic elements. Here the results indicate that VMI's higher level routines and processes have been optimised by the application of GUSI, namely the GS1 XML electronic messaging and standard message choreography. In Novozymes' experience, the services from GS1 and CGF have resulted in a robust business case with high buy-in potential for trading partners.

The lesson from the case study is that a holistic approach is required when adding value in the supply chain. When crafting a strategy to achieve this, components of advantage should be seen as a synergistic set of competencies rather than isolated resources. Superior integration may give an organisation more operational efficiencies, but unless it is blended with improved processes, such as VMI, optimal performance may not be achieved. Whilst innovative product development is a key driver for Novozymes sustainable growth and profitability, logistical arrangements also make a significant contribution in adding value to Novozymes' overall profitability.

By eliminating inefficiencies in the supply chain Novozymes logistics and supply chain management department is actively contributing to the current and future success of their organisation, partially enabled through the use of GS1 standards and GUSI harmonised processes.

5.3 Implications for Practitioners

In general, it is Novozymes' experience that GUSI VMI leads to closer collaboration between the selling and the buying companies. Handing over the replenishment responsibility to a supplier takes collaboration in a dyadic trading partner relationship to the next level which requires a high level of trust and dedication from both sides. In general, a close trading partner relationship is built up over the years and not necessarily an outcome resulting solely from a GUSI VMI implementation, however, GUSI VMI is definitely part of a bigger 'package' of collaborative services which the professional supplier must offer to highly valued customers to strengthen the partner relationship.

This study demonstrates that it is possible to improve integration capabilities by applying global upstream standards. In this case, a reduction in time effort of around 60% was reported by Novozymes in total system integration with their trading partners. Additionally, by optimising IT integration, trading partners are on-boarded faster than in a non-GUSI environment with the result that it is possible to make use of the VMI process five months sooner than in a traditional non-GUSI environment.

The true gains within Novozymes are, however, not the reduction in transaction costs and quicker on-boarding of new suppliers, but the buyer being integrated with the product that Novozymes produce; the more efficient the supply of product, the leaner and more cost effective the supply chain. So with this in mind it could be said that processes and procurement scenarios (e.g. TOM & VMI) are enabled more quickly when deployed using foundational eBusiness standards. These types of additional higher rent generating processes have, in Novozymes' case, been seen to improve the demand transparency.

5.4 GUSI, a Normative Prescription for Supply Chain Management

Is there a normative prescription to a successful supply chain scenario? This case study suggests that the application of GUSI reduces internal IT integration effort and plays a role in minimising transaction cost economics (TCE) through reduced IT integration resource allocation. Theory claims that the exchange of demand information from the customer to the supplier (short term material requirements and long term forecast), will reduce the total amount of inventory, and hereby working capital, in the supply chain (Disney & Towill, 2003; H. L Lee, et al., 1997; Hau L. Lee, et al., 1997). This is often referred to as the so-called bullwhip effect. With some customers, it has been possible to reduce inventory levels on both the sides, but due to the very long production lead times, accuracy in long term forecasting from

the customer is crucial to achieve this benefit.

However, when studying the case through the resource based view (RBV) lens, the benefits can be more far reaching than simply reducing transaction costs of technical effort and allocation of in-house resource. Whilst transaction costs can be significant and are frequently referred to synonymously with efficiency (Williamson, 1981), it is often the allocation and configuration of bundled resources that provides organisations with greater benefits, vis-à-vis the competition (Wernerfelt, 1984). Of course, at some level, companies diverge process and strategy to maintain, or develop, competitive advantage and the concept of applying a normative prescription to attain advantage may seem counter intuitive to this goal.

This idea can be discussed further by developing a chemistry metaphor. Two chemical compounds may contain the same elements in the same proportion, yet because their atoms are linked together differently, the identities of the substances are distinct (Atkins, 1991). This is the essence of the advantage which Novozymes has created, where the advantage is in the quality of 'the melt'. Whilst individual components of the GUSI VMI implementation are not unique, the aggregation of capabilities adds to competitiveness.

Benefits in a GUSI VMI environment could be summarised as the addition of bundled organisational capabilities, namely, processes, message choreography and master data alignment, all offering the ability to better manage product demand data from the downstream customer. This targeted transparency management is predominantly where demand is met with as little demand distortion between the two parties as possible enabled by efficient and effective business information exchange. This added business value is the ultimate goal of an efficient supply chain and the GS1, CGF, GUSI initiative has been seen by Novozymes as a factor that facilitates the implementation of higher rent generating processes.

6. DIRECTION FOR FUTURE RESEARCH

6.1 Risk Reduction

The case study brings to light the role of the standards organisation as a neutral broker in business standards, specifically their governance, maintenance and development. It could be said that the value of a standard increases proportionally to the amount of users it attracts, so, it is of some importance to the manager to ensure that any standards which are adopted should be taken up by as many stakeholders as possible to increase the value of the investment. Further studies could review the effect standards organisations have on risk reduction in the areas of adoption and standards maintenance, an important aspect if a community of users want neutral support and governance of foundational standards.

6.2 Master Data

Whilst the GUSI XML business messaging standard prescribes GS1 keys both as the unique identification of; locations, routing endpoints, business roles and unique product identification, the benefits of using GS1 keys is not in the scope of this case study. CGF has made a good attempt to demonstrate the value add of the GS1 master data key, the Global trade Item Number (GTIN) (2007) , but a fuller, academic and open study could open up these benefits to a wider audience. The application of standardised identification is a pre-requisite for Master Data Alignment [product and location] (MDA) and it would be a natural progression in the research to ascertain how GUSI business standards' contribute to the benefits associated with MDA.

6.3 Benchmarking GUSI

It would be of academic, strategic and operational interest to measure results form a specific application of GUSI against those of another organisation or sector benchmark to measure the actual benefit gained from the GUSI implementation. Any new research following this path could be used to correlate with the Novozymes' results and add validation to the adoption potential of GUSI standards.

ACKNOWLEDGEMENTS

The authors gratefully acknowledge the co-operation of Novozymes A/S in the creation of this paper and are grateful for the review comments received from Regenald Kramer from GS1 Global Office.

REFERENCES

Akkermans, H., Bogerd, P., & van Doremalen, J. (2004). Travail, transparency and trust: A case study of computer-supported collaborative supply chain planning in high-tech electronics. *European Journal of Operational Research,,* 153(2), 445-456.

Angeles, R., Corritore, C. L., Basu, S. C., & Nath, R. (2001). Success factors for domestic and international electronic data interchange (EDI) implementation for US firms. [doi: DOI: 10.1016/S0268-4012(01)00028-7]. *International Journal of Information Management, 21*(5), 329-347.

Atkins, P. W. (1991). *Atoms, Electrons, And Change.* New York, USA.: Scientific American Library.

Barratt, M., & Oliveira, A. (2001). Exploring the experiences of collaborative planning initiatives. *International Journal of Physical Distribution & Logistics Management, 31*(4), 266-289.

BERR. (2009). The globalisation of value chains and the industrial transformation in the UK *BERR Economics papers* (pp. 1-110): Department for Business Enterprise & Regulatory Reform.

Butner, K. (2010). The smarter supply chain of the future. *Strategy and leadership, 38*(1), 22-31.

CGF. (2010, 2009). What is The Consumer Goods Forum? Retrieved 19 August, 2010, from http://www.theconsumer-goodsforum.com/

Christopher, M. L. (1992). *Logistics and Supply Chain Management.* London: Pitman Publishing.

den Butter, A. G., & Linse, A. (2008). Rethinking Procurement in the Era of Globalization. *MIT Sloan Management Review, 50*(1), 76-80.

Disney, S. M., & Towill, D. R. (2003). The effect of vendor managed inventory (VMI) dynamics on the Bullwhip Effect in supply chains. *International Journal of Production Economics, 85*(2), 199-215. doi: Doi: 10.1016/s0925-5273(03)00110-5

Duque, J. F. Z., & Skøtt, H. (2009). *Novozymes GUSI benefit analysis.* Internal analysis. Novozymes. Copenhagen.

EC. (2010a). Europe 2020: a new economic strategy Retrieved 5 March, 2010, from http://ec.europa.eu/eu2020/

EC. (2010b). Europe 2020: Commission proposes new economic strategy in Europe. . *Europa, Press relase RAPID* Retrieved 5 March, 2010, from http://europa.eu/rapid/pressReleasesAction.do?reference=IP/10/225&format=HTML&aged=0&language=EN&guiLanguage=en

Europe's Information Society. (2010). i2010 benchmarking framework. *Benchmarking* Retrieved 13 February, 2010

Fabbe-Costes, N., Jahre, M., & Rouquet, A. (2006). Interacting standards: A basic element in logisitc networks. *Journal of Physical Distribution and Logisitcs Management, 36*(2).

GCI. (2006). Business Case Outline & Key Success factors for Implementing GUSI. In G. C. Initiative (Ed.), (Vol. 1). Cologne.

GCI. (2007). GUSI reasserts GTIN and GLN value for the Upstream Industry Model (August 2007 ed.). Cologne: GCI/Consumer Goods Forum.

GCI. (2008). About GCI Retrieved 3 June, 2010, from http://www.gci-net.org/e2/

GCI. (2010). Global Upstream Supply Initiative Retrieved 30 March, 2010, from http://www.gci-net.org/e8/e77/

Ghemawat, P. (2006). Apocalypse Now? [Article]. *Harvard Business Review, 84*(10), 32-32.

Ghemawat, P., & Hout, T. (2008). Tomorrow's Global Giants. [Article]. *Harvard Business Review, 86*(11), 80-88.

GS1 in Europe. (2010). GS1 in Europe: Harmonised standards and solutions for European Business Retrieved 3 June, 2010, from http://www.gs1.eu/?content=20

Halldorsson, A., Kotzab, H., Mikkola, J. H., & Skjøtt-Larsen, T. (2007). Complementary theories to supply chain management. *Supply Chain Management: An International Journal, 12*(4), 284-296. doi: 10.1108/13598540710759808

Halldorsson, A., Larsen, P., & Poist, R. (2008). Supply Chain Management: A comparison of Scandinavian and American perspectives. *International Journal of Physical Distribution and Logistics Management, 38*(2), 126.

Hertz, S., & Hultman, J. (2008). On Global Supply Chain Development. In J. S. H. Arlbjørn, A; Jahre, M; Spens, K (Ed.), *Northern Lights In logisitcs and Supply Chain Management* (1 ed., pp. 267-268). Copenhagen: Copenhagen Business School Press.

Hill, D. S. (2009). The UN/CEFACT Business Messaging Standard: A Potential Source of Competitive Advantage. *Journal of Operations and Supply Chain Mangament, 2*(1), 46-60.

Hill, D. S., & Duque, J. Z. (2008, October 28). Global upstream standards in the pharmaceutical sector. *På Stregen, 5.*

Hill, D. S., & Oscarsson, J. (2010). [Carlsberg's application of GUSI].

Holmström, J. (1998). Business process innovation in the supply chain – a case study of implementing vendor managed inventory. *European Journal of Purchasing & Supply Management, 4*(2-3), 127-131.

Hsieh, C., & Lin, B. (2004). Impact of standardisation on EDI in B2B development. *Industrial Management & Data Systems, 104*(1), 68-77. doi: 10.1108/02635570410514106

IBM. (2009). The Smarter Supply Chain Of The Future. Somers.

Iskanius, P., & Kilpala, H. (2006). One step closer towards e-business—the implementation of a supporting ICT system. *International Journal of Logistics Research and Applications, 9*(3), 283-293. doi: 10.1080/13675560600859482

Lambert, D. M., Cooper, M. C., & Pagh, J. D. (1998). Supply Chain management: Implementation Issues and Research Opportunities. *The International Journal of Logistic Management, 9*(2), 1-19.

Lee, H. L., Padmanabhan, V., & Whang, S. (1997). The Bullwhip Effect in Supply Chains. *Sloan Management Review,* 933-132.

Lee, H. L., Padmanabhan, V., & Whang, S. (1997). Information Distortion in a Supply Chain: The Bullwhip Effect. *Management Science, 43*(4), 546-558.

Lee, S., & Lim, G. G. (2005). The impact of partnership attributes on EDI implementation success. *Information & Management, 42*(4), 503-516. doi: DOI: 10.1016/j.im.2003.03.001

Lehtonen, J., Småros, J., & Holström, J. (2005). The effect of demand visibility in product introductions. *International Journal of Physical Distribution & Logistics Management, 35*(2), 101-115.

Lessard, D. (2008). Global Strategy and Organization. *MIT Open Courses* Retrieved 1st June, 2010, from http://ocw.mit.edu/courses/#sloan-school-of-management

Liedka, J. (1999). Linking competitive advantage with communities of practice. *Journal of Management Inquiry, 8*(1), 5-16.

Millen, R. A., & Ukena, J. (1995). EDI usage in the motor carrier industry. *International Journal of Physical Distribution & Logistics Management, 25*(6), 23-40.

Porter, M. E. (2001). Strategy and the Internet. [Article]. *Harvard Business Review, 79*(3), 62-78.

Porter, M. E. (2001). Strategy and the Internet *Advances in Strategy* (pp. 28). Boston, MA: Harvard Business School Publication.

Porter, M. E. (2008). The five competitive forces that shape strategy. [Article]. *Harvard Business Review, 86*(1), 78-93.

Rodon, J., Ramis-Pujol, J., & Christiaanse, E. (2007). A process-stakeholder analysis of B2B industry standardisation. *Journal of Enterprise Information Management, 20*(1), 83-95.

Shibata, M., Ariga, R., Suga, S., & Sato, A. (2007, 19 April). *GCI Japan & Japan's EDI-XML pilot.* Paper presented at the GS1 Global Standards Management Process, Sofitel, Budapest, Hungary.

Shpilberg, D., Berez, S., Puryear, R., & Shah, S. (2007). Avoiding the Alignment Trap in IT. *MIT Sloan Management Review, 49*(1), 51-58.

Skjoett-Larsen, T., Thernøe, C., & Andresen, C. (2003). Supply chain collaboration: Theoretical perspectives and empirical evidence. *International Journal of Physical Distribution & Logistics Management, 33*(6), 531-549.

Småros, J. (2003). Collabaorative Forecasting: A Selection of Practical Approaches. *International Journal of Logistics: Research and Applications, 6*(4), 245-258. doi: 10.1080/13675560310001626981

Smith, A. (2003). Exploring potential strategic impacts of XML-related technologies. *Information Management & Computer Security, 11*, 98.

VICS. (2010). Collaborative Planning, Forecasting & Replenishment (CPFR®) Committee Retrieved 3 June, 2010, from http://www.vics.org/committees/cpfr/

Wernerfelt, B. (1984). A Resource-Based View of the Firm. *Strategic management journal, 5*(2), 171-180.

Williamson, O., E. (1981). The Economics of Organization: The Transaction Cost Approach. *The American Journal of Sociology, 87*(3), 548-577.

WTO. (2008). World trade 2008, Prospects for 2009 Retrieved 22 January, 2010, from http://www.wto.org/english/news_e/pres09_e/pr554_e.htm

AUTHOR'S BIOGRAPHY

Douglas S. Hill is the Chief Operating Officer (eBusiness) at GS1 Denmark and an MBA graduate from the University of Liverpool, UK. Douglas is currently a PhD Management student studying with the Supply Chain Research Group at The University of Southampton, UK. His research interests are specifically in the application of technology as a driver for competitive advantage, transparency, neural supply networks and inter-organisational relationships.

Juan Francisco Zurita Duque holds a M.Sc. from Danmarks Tekniske Universitet. He is currently Team Lead for SAP Solutions and Integration at Novozymes A/S (DK). His area of professional specialization is in the line of Enterprise Service Oriented Architecture, Business Process Integration and Integration Standards.

Helle Skøtt has a M.Sc. in Supply Chain Management (1996) and manages supply chain integration for Novozymes A/S. Helle is a part time lecturer on Supply chain management at Copenhagen Business School, in Denmark.

2

The adoption of Global Sourcing by a Brazilian Company

Moema Pereira Nunes
Unisinos
moemanunes@hotmail.com

Luciana Marques Vieira
Unisinos
lmvieira@unisinos.br

José Antônio Valle Antunes Jr.
Unisinos
junico@produttare.com.br

ABSTRACT: Companies are going abroad looking for advantages from alternative sources of supply. Primarily associated with the lack of local supplier or lower acquisitions costs, these reactive reasons have driven business to a more proactive perspective by the adoption of a Global Sourcing (GS). Companies from emerging countries are developing their own GS, just like multinational from developed countries. These "late movers" have a strategic motivation, and goals, to internationalize their activities. This paper reviews and integrates literature from Operations Management, Strategic Management and International Business areas to develop a framework that poses 13 theoretical propositions. Then, the paper presents a case study carry out in a Brazilian electro- electronics company that develops and manufactures its products overseas and distributes them in Brazil and other South American markets. This company replicates the strategy developed in its home country and spread value chain activities through global suppliers, especially those located in Asia. Findings reveal that the company has the capability to analyze its value chain to focus on the activities that add more value: those are product development and distribution in South America. Each proposition of the framework was illustrated by the empirical evidence from the case study. The findings provide insights to guide further research on the adoption of GS by companies from emerging countries. Some propositions of the theoretical framework were not found due to the size and strategy of the analyzed company and new propositions were developed based on the empirical findings for further research.

Key words: Supply Chain. Global Sourcing. Electro- electronics industry.

1. INTRODUCTION

The search for alternative sources of supply in foreign markets is considered an opportunity for competitive advantage seeking to companies. Primarily associated with the lack of local suppliers (of products, services and technology), or lower acquisitions costs, these reactive reasons have driven business to a more proactive perspective (Monczka and Trent, 1991; Bozarth, Handfield and Das, 1998; Harris, 2006; Servais, 2007; Dutton, 2008). The adoption of strategic approach to sourcing globally became a currently practice in order to achieve more competitiveness with their supply base. Instead of just purchasing abroad, Global Sourcing (GS) means a strategic orientation for the search and monitoring of global suppliers and an efficient management by integrating and coordinating activities related to the functional areas as well as units of local purchases.

The inclusion of GS in a company's strategy can be considered as a recent phenomenon in some countries. Even with companies that are used to making international purchases, the consolidation of this activity into their strategic plan with long-term commitment is increasing. This situation leads to some questions that the companies will face during the adoption of GS. Theories can be developed from the experience of companies in other countries, and also based on companies' performing in different business environments. The economies of countries such as Brazil, Russia, India and China are growing fast in a moment when they have more access to technology and communication facilities. The establishment of complex governmental agreements can both facilitate and impose barriers to international trade. Companies from emerging countries are presenting a fast internationalization process, characterized by bold and aggressive methods in the early stages of the process (Sirkin et al, 2008). At the same time, the competitive advantages of emerging companies are usually related to price competition, which is more difficult to sustain than technological capacity or branding (Gammeltof, Barnard and Madhock, 2010). Considering this argument, it is possible to highlight that we may have a background that support the assumption that the GS adopted by emerging companies may differ from the strategy adopted by companies from developed countries.

The surge of MNCs from emerging countries is reshaping the structure of international business (Gammeltof, Barnard and Madhock, 2010); these types of companies accounted for approximately one-quarter of all MNC parent companies globally (Tolentino, 2010). This paper represents an effort to expand the knowledge based on the experience of a company from an emerging country. This will provide new questions to further research in this area of knowledge.

Brazilian companies have grown in the last years in terms of volume of international operations. This rapid growth has been surprising researchers, politicians, and even managers. This is leading to the investigation of Brazilian companies strategies in an effort to capture the characteristics of these new MNCs (Fleury, Fleury and Reis, 2010). Fleury, Fleury and Reis (2010) find the employed management models of these companies, which are based on a combination of organizational skills and management practices, to be noteworthy in the internationalization of Brazilian MNCs. Examining Brazilian MNCs, these authors found that whereas the internationalization of traditional MNCs took place through seeking new markets and access to resources, emerging MNCs from Brazil are engaging a mix of activities that also includes searching for strategic assets and enhancing efficiency. The motivations and goals of these "late movers" in terms of internationalization are, therefore, more strategic.

The new competitiveness pattern that emerge from new challenges bring forth the need to investigate how the previously knowledge developed about GS can explain the adoption of this strategy in emerging countries. As Brazil is one of the emerging countries whose companies have establish new patterns of internationalization this paper seeks to investigate: *"How the adoption of a global sourcing strategy is being conducted by Brazilian companies?*

Trent and Monczka (2003) highlight the need for future research on robust GS processes because they found diversity in terms of the development and implementation of this strategy in the companies they investigated. Samli, Browning and Busbia (1998) reveal the need for studies that investigate the adoption of GS, including business involvement in this process and the incorporation of this activity into corporate strategic planning. By investigating emerging issues in supply, Sheth and Sharma (1997) determined that GS activities should be explored further because of the opportunity to obtain a competitive advantage through this strategy. At the same time, the authors highlight that cultural and legal differences among countries are critical factors that are directly relate to GS.

The literature review is divided in five sub-sections: antecedents to GS, opportunities to GS, process of GS, difficulties and challenges, and GS results. In the sequence, the method of the study is described followed by the case description and analysis. The last section contains the final considerations of the study.

2. ANTECEDENTS OF GS

Companies purchase abroad in order to reduce costs and to access other advantages from the supplier country or the supplier company. Considering the Transaction Costs Economics Theory, the essence to purchase is the complex cost analysis and the evolution of the research related to sourcing are showing that even cost is the essence, other benefits are being achieving through this activity.

Investigating the advantages from GS, Alguire, Frear and Metcalf (1994) identified that companies can obtain comparative and competitive advantages through this strategy. It is the possibility of advantages that lead companies to source globally. It is the possibility to increase the advantages that will lead them to adopt GS. This evidence leads us to the investigation of the strategic orientation of the companies to adopt GS.

Strategic orientation

The motivations to GS can be related with the company's products or input's features. For example when a product required a specific raw material that cannot be made by the company. Another situation can be relate with the supplier, like when one can obtain a product just in a specific country, or can get it with a less total cost. These motivations can be divided in two types of advantages. The comparative advantage's motivations are those that leads the company to capture local cost advantages though the utilization of foreign suppliers. The competitive advantages are related to the company's ability to offset competitive disadvantages other than costs (comparative advantages), like access new technologies, delivery improvement, flexibility to change the input's features, product reliability improvement, quality improvement, quality control improvement, establishment of alternative supply sources, increase in the number of available supplier, supplier reliability improvement, access advantages from supply's market, access advantages from supply's core competence, anticipate material needs to new products in development, anticipate materials needs in

case of demand changes, better negotiations conditions, introduce of competition on the supplier base, reduction of product development cycle, customer service improvement, establishment of presence in global market, get the opportunity to sale to a specific market or country, meet supply constraints imposed by government, offer global support to local products, react to competitor's practices, and support to the company own international operations (Monczka and Trent, 1991; Bozarth, Handfield and Das, 1998. Dornier et al., 2000; Cho and Kang, 2001; Christopher, 2002; Jin, 2004; Agndal, 2006; Harris, 2006; Knudsen and Servais, 2007; and Dutton, 2008).

P1: The adoption of GS is motivated by competitive advantages.

Organization Structure

The structuring of a company's resources, process and supplier management are some of the aspects that must be included in the GS management process. The analysis of the structure will enable us to know if the company is purchasing globally with a strategy view and the process enables us to know how the company is doing the sourcing (Lima, 2004). In GS studies, one must consider that an activity can be developed either by a single company that operates in the global marketplace or by a set of companies belonging to a corporation with units in a single country or different locations. When the object of study is the second group of companies, some dimensions are add to the management of this activity. Much of the discussion about GS is around centralization versus decentralization of international purchasing (Arnold, 1989, Monczka and Trent, 1991; Trent and Monczka, 1998, 2003, 2003a; Arnold, 1999; Trautmann, Bals and Hartmann, 2009). According to Porter (1986a), an internationalization strategy presents two key-dimensions. The first is the localization of each company's value chain activity worldwide (configuration). The second refers to the way these activities will be coordinated (coordination). Considering global companies, the success in adopting GS implies the delegation of a sufficient degree of autonomy of its subsidiaries. Although the corporation must follow a homogeneous orientation, this flexibility will allow greater speed of action in the market. The consequence is the need for better coordination between units of the same corporation (Arnold, 1989). Extending this discussion about the degree of centralization needed in managing GS, Matthyssens and Faes (apud Arnold, 1999) present arguments for and against the autonomy of the units. Gelderman and Semeijn (2006) affirm that the necessity to coordinate these activities emerges when considering that a company will perform purchasing activities on a local and global basis. The aspects of formalization became important to the management of GS when dealing with companies with different subsidiaries and different purchase units (Hartmann, Trautmann and Jahn, 2008).

P2: The centralization of GS activities is related with the potential synergy between units, and their supply needs, that will results in better sourcing conditions.

Looking at the company configuration to implement GS, it is possible to affirm that there are some requirements to implement it. To Arnold (1989), the first is the company's size. At the same time that big companies seem to have more resources available, the small usually are more predisposed to flexibility. According to Trent and Monczka (2003), companies that engage in GS are larger and more likely to have competitors that are multi-regional or global than comparing to companies that make international purchases. Knudsen and Servais (2007) say that the internationalization of purchase activities require resources and time to be developed. Observing the international environment for small and medium enterprises, the authors argue that the size limits their internationalization expansion but the experience in international purchases can facilitate this process. A second aspect is the inferior position given to the purchase area in organization. It can limit the ability to take advantage of opportunities due to internal neglect. This aspect is supports by Arnold (1989) and Quintens, Pauwels and Matthyssens (2006), to whom the top management supports is a facilitator to GS. This will not necessarily lead to (more) global purchasing, but ease its implementation. Besides that, the internal articulation between areas requires balance between the areas to which the collaborative process can be accomplish. The identification if GS reflects a strategic approach to procurement in the company implies in the research of the corporate levels that are involved with the purchase decision process within the company's structure. GS requires that these decisions be taken by top-managers and in the case of corporations, the corporate direction and not isolated units. This level of decision-making seems to be crucial in order to give the necessary importance to purchase function within the organization while aligning strategic planning and the planning of the purchase function.

P3: The adoption of GS implies that the availability of resources for establishing and managing the activity is relative to the organization's size and, the importance attributed to GS, including the top management support and the industry features.

A last aspect related to company's structure is the presence of the purchaser at the supplier country. The internationalization of supply with the presence of the purchaser company in supplier's country can happen in different ways that represent a continuum of involvement (Harris, 2006). The first approach is the use of trading companies to broker the negotiations and activities between purchaser(s) and supplier(s). Considering that they are subcontractors, it does not imply investments of the purchasing company. Moreover, as intermediaries, they usually conduct all or part of the operational activities. The second way is the use of local representatives of the purchasing company in supplier markets. One form of representation entails the opening of International Purchase Offices – IPO. The establishment of a business unit abroad and assign the same responsibility on the development of a product, regional or global, is another strategy observed by Harris (2006). This way the company can capture the best opportunities available in a particular place, related technology and production.

P4: The presence of the purchase company in the supply country motivates the adoption of GS.

3. OPPORTUNITIES

To obtain higher gains from GS, it is prior to get involved in the identification of opportunities with global suppliers. The identification of the real potential of a supplier requires a joint analysis of purchases with other functional areas such as product development (Quintens, Pauwels and Matthyssens, 2006). The logistics also become important to analyze the operational feasibility of the acquisition process. The relationship with these and other areas is evident, therefore, as a dimension to be investigate in GS. With this research, one must analyze the mechanisms of integration used; how this integration occurs and how activities are coordinated among functional areas of business to ensure transparency and speed in these interactions.

To be able to identify opportunities in the global world, companies must dedicate resources to the sourcing area with the focus on the analysis of internal and external opportunities. GS requires the monitoring of actual and potential suppliers' environments, including the investigation of macro and micro-economic variables in order to identify the sourcing opportunity. Part of this work is related with the establishment of the alignment of internal functions and activities, and the investigation of join sourcing opportunities inside the company's structure. The proximity of purchase area with others inside the company results in the identification of potential supply demands, facilitating the pro-active approach of sourcing area to search potential supplier markets and companies.

P5: The supply opportunity analysis' process includes the investigation of the supplier company, the inputs, and the supply and sourcing environments, as well as customer requirements improving the adopting of GS.

4. PROCESS OF GS

The GS process must be view as possible more complex process to the purchasing area to promote the entrance of inputs in the materials sourcing flow. It must be consider that this area is not responsible just for the supply of materials, as services are also related. The investigation of a sourcing process involves the study of (1) the selection and development suppliers and (2) the analysis of the participation of the purchase areas in the supply materials flow (Lima, 2004).

Supplier's Management

GS assigns responsibility to the company regarding the search for potential suppliers and the development of relationships with them. This search includes everything from identifying potential market suppliers to the selection of specific supplier for a component (Trunick, 2006). To Trent and Monczka (1998), the tendency to concentrate the company's purchases generates an expansion of the need for supplier development. Traditional supplier management models highlight the involvement of suppliers in the business, however, they do not make the distinction between local and global (Grieco, 1995, Cebi and Bayraktar, 2003, Kamath and Liker, 1994, Simchi-Levi, Kaminsky and Simchi-Levi, 2003). Looking at the GS approach, the involvement of suppliers into new product development is an unexplored aspect. Looking at the GS approach, the involvement of suppliers into new product development is an unexplored aspect. Considering that GS leads to a close relation between purchasers and suppliers, and with the areas inside the company, and that the motivations include the source for new technologies

and access to supply capabilities, it is reasonable to question how is the engagement with global suppliers with respect to new product development.

P6: The adoption of GS is related with the involvement of suppliers into new product development as the units of the company and units areas are more related, but on the other hand work with global suppliers is more difficult comparing with local suppliers.

Investigating the relationship of Finnish companies with Chinese supplier, Salmi (2006) found that these relationships are built gradually. Time and commitment of the companies are important factors for the establishment of a relationship. According to Knudsen and Servais (2007), companies involved in international transactions have greater concern about the suppliers' monitoring to avoid the increase of total costs and the decrease of product's quality. Sometimes, the relationship can be better developed in contracts with innovate suppliers, which are more favorable to flexibility (Dornier et al, 2000). Investigating the relationship of four suppliers with a purchaser company, Ghauri, Tarnovskayaand and Elg (2008) identified the importance attached to developing interpersonal relationships with transparency. They also highlighted that technological and financial support are ways to develop the relationship; that people turnover leads to loss of efficiency in the process. According to Bozarth, Handfield and Das (1998), despite the recognition of the importance of pro-active international suppliers, there are few efforts that identify the management of international suppliers through a strategic view.

P7: The choice to use foreign suppliers based on the purchaser country will be related with support, customer service, and cultural aspects related to the supplier management process.

Purchase process

The relationship between different areas inside each company and different units of the company requires a standardization of materials, maybe through codification, in order to facilitate the communication flows. It is prior that the importance of each material also be consider in GS and to do so, the use of materials or purchasing portfolios represents an opportunity to a better management process. Different studies have revealed the benefits of its use and different variables were presented in the portfolios such as Gelderman and Semeijn (2006), Smith (1999) and Trautmann, Balls and Hartmann (2009).

P8: The GS activity is oriented to inputs consider strategic for the company, and the other inputs should be included in the context of this strategy.

There is a need for qualified personnel working with GS activities. Trent and Monczka (2003a) identify that those professionals with knowledge and skills are the most important success factor for GS. These professionals must be able to make presentations and communicate effectively, to think holistically through a company or region, and to work in an environment of cultural diversity. Mulani (2008) recognizes that the focus on people contributes to success in GS. This focus can be perceive through the training programs, educational opportunities, the offer of attractive benefits packages, and the forecasting of international work opportunities, mainly focusing on the maintenance of people on the team.

P9: The GS approach requires qualified personnel and continuously training of them in order to identify better opportunities and conduct efficient sourcing process.

The availability of communication tools is another factor to be consider in a company's structure. The success of GS depends on a high level of transparency that leads to the pro-active cooperation between the members of the supply chain, the identification of what is happening and the management of situations that are inconsistent with the initial planning (Wilding and Braithwaite, 2007). Wilding and Braithwaite (2007) highlight that a company needs capabilities related to communication and information flow in order to implement GS. Information, considered as the ability to request information and data, and as the ability to identify common requests, is one of the factors for the international supply activity; this is perceive at companies that adopt GS as they rely on a wider array of communications tools (Trent and Monczka, 2003).

P10: The existence of communication tools and platforms are important to global supply management, including actual and potential suppliers, and actual and potential demands.

5. DIFFICULTIES AND RISKS

Through the literature review, one can identify that the risks from GS to the purchase company include the possibility of a decrease in the company's agility and flexibility, the increase of distance, cost and the number of intermediaries in the supply chain, the maintenance of the analysis' focus in specific source

operation instead of the complete process what reduce the ability to analyze the situation, the possibility of an increase in the total costs, the failure of logistics support, difficulties to deal with cultural differences, regulations and country uncertainty (Levy, 1995, Bozarth, Handfield and Das, 1998, Cho and Kang, 2001, Zeng and Rosseti, 2003, Christopher, Peck and Towill, 2006, Butter and Linse, 2008, Steinle and Schiele, 2008).

P11: GS includes the management of risks, considering cultural and governmental diversities and its special characteristics.

Barriers can be seen as factors that make it more difficult or even impossible to pursue or intensify GS. Pauwels and Matthyssens (2006) present a set of barriers divided in five categories: product, company/ management, network industry /competition and environment. Another approach to understand the barriers to GS was developed by Alguire, Frear and Metcalf (1994), dividing them into internal and external.

P12: The analysis of external barriers is important during the process of purchase opportunity analysis to avoid risks and ensure benefits.

6. GS RESULTS

Comparing with companies that make international purchases, companies adopting GS can better understand that there are many business opportunities beyond what is being purchase. These companies realize better performance and costs reduction. To them, performance improvement and cost reduction opportunities are more widely available from their sourcing efforts. They can make changes in the supply items more quickly, lead, and coordinate strategic reviews more regularly in order to promote consistency by creating a common language and approach of searching for suppliers at the organizational level (Trent and Monczka, 2003). They are also able to perceive their strategy implementation progress to be further along, face more rapid changes to product and process technology, and rate key aspects of their sourcing process as more similar across geographic locations and buying units (Trent and Monczka, 2003). As Mulani (2008) emphasis, supplier involvement is a possibility to absorb knowledge, leverage capabilities, maximize contracts, and continually reduce total costs.

P13: The adoption of GS leads to competitive advantage comparing with companies that purchase internationally.

Through a literature review, a set of propositions is developed for each 5 dimensions from the theoretical framework. They present a set of interconnected activities that, together, represents the adoption of GS by companies, and it will be investigate in a case study. Even though this research is focusing on an emerging company from Brazil that is not a MNC, the framework was developed in a broader context in order to comprehend the whole process in more complex environment of MNCs. In an effort to simplify the analysis of the framework, five dimensions were identified in the theoretical framework: (1) antecedents of GS, including strategic orientation and organizational structure, (2) opportunities,(3) process of GS, including supplier's management and purchase process, (4) difficulties and risks, and (5) results.

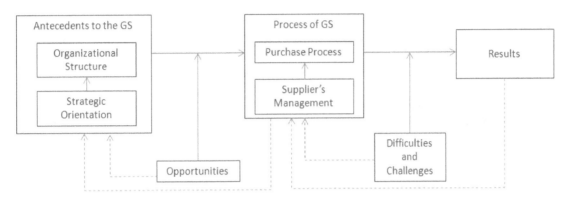

Figure 1 – Theoretical Framework

7. METHOD

In order to achieve the objective of this paper, a case research methodology was used as the research is related with the investigation of a phenomenon in a different context and on the exploration and better understanding of an emerging and contemporary phenomenon (Barrat, Choi and Li, 2011). The theoretical framework was used to conduct the exploratory case study and new perspectives were identified as a result of this analysis.

According to Yin (2001), a case study is a strategy that allows the research to investigate the phenomenon within its real-life context, especially when the boundaries between phenomenon and context are not clearly defined. The case study developed here can be classified as a deductive study as it had the purpose to test existing theory in line with other studies in the general business disciplines, like management. Barrat, Choi and Li (2011) identified that the use of a deductive purpose to develop a case study is being used by researchers in a small scale comparing with inductive purposes. In a deductive case-based study, the researchers are looking for patterns in the emerging data to compare with the theoretical derived propositions (Barrat, Choi and Li, 2011). Johnston et al. (1999) highlight that there are three main requirements for use deductive qualitative research methodology: (1) the case study must begin with an existing theory for the development of research propositions; (2) a systematic and logical research design should be followed; and (3) researchers should implement evaluation criteria to independently assess potential biases and to ensure the methodological rigor.

The first approach of the researcher to the industry sector occurred on August 2011 in a meeting with representatives of the industry and the state of Rio Grande do Sul government, when the industry association presented their interest in developing improvements in the global sourcing process of their associate companies. A second meeting occurred in September 2011, to discuss the research objectives to the local Director of the association and two advisors. Ideas on their global sourcing project were also presented and a potential company was selected to be the first investigated case. Data collection included a direct observation and in-site interview with the owner of the company who is also the responsible for the global sourcing process on October 2011. This interview was conduct by a set of questions developed based on the theoretical framework dimensions. The interview lasted three hours and the audio was record. The data were submit to a content analysis with the support of NVivo.

The codification process was conducted at NVivo based on the propositions. It was identified a set of important aspects that must be analyzed according to each propositions and in the transcription of the interview. Even though the questions were separated according to the propositions, during the interviews, the information reported to each question could be related with others propositions and this codification process was important to assure the best use of the information. The transcription was read and analyzed three times in order to assure that the codification process was capturing all the information. After that the codification report with the text separated by nodes were analyzed and some adjustments were made in order to assure the consistence of the information generate by the data. The use of NVivo was important for the research in order to organize the information and assure the correct use of them in the research process. The data base that was developed with the software can be considered a way to reduce the fragility of the case-based study related with the fact that the data are analyzed in an intuitive way in a qualitative study. The preliminary results were present to one of the industry association advisor in order to assure that the investigated company represent the industry sector as an exemplary case. This conversation lasted one hour. It was important as a way to validate the results.

8. DESCRIPTION OF THE CASE

Company A is a Brazilian owned business of the electro-electronic sector with 3 decades of existence. The company's sales in 2010 were in between R$ 20 – 50 Millions (10-25 Millions U$D). The production is centralized in one processing plant in Brazil, even though some products are purchased finished from Asia. Related with the sourcing process, the participation of international suppliers is increasing in the last five years, from 2% in 2007 to 20% in 2010. The internationalization of Company A is also reflected by its exports increase that is mainly to South American countries.

The foundation of the Company A was related with an innovative product developed by their owners. The search for innovation was not just related with product, but also with business model. The Company is focus on its core competence. Its products are more expensive than Asian similar ones. To be competitive, the company focuses on customized products, even those sold under the customer's brand. It has decentralized the manufacturing process with the engagement of a group of suppliers located closer to the company's site. This relationship with suppliers is replicated currently with global suppliers. The Company A used to manufacture its products in suppliers units, and is involved in carry out the quality control, packing, expedition and shipping at home. Inside Brazil, the company is changing this strategy because of logistics cost, but the relationship with suppliers is still an important aspect of company's strategy. The Company realized that it is more than a manufacturing company and has been focusing on a distribution process that must be used with complementary products. These products are sourced in international markets and represent 20% of the company's sales.

The approach to international suppliers became more than 10 years ago with a first business mission promoted by the industry association (ABINEE – Brazilian association of electro- electronics industry). Since then, the company managers keep travelling to Asia every year to visit business fairs and suppliers' units. Potential suppliers are also identified in international fairs that are organize in Brazil when international suppliers came to exhibit or to visit.

The adoption of a strategic view of sourcing was an outcome of the Company's growth. As it moves from being a small local company to a middle size company, the managers realized that the benefits offered to small companies were reduced. At the same time, the established competitors started to treat the Company as an equal and the consequence could be the failure or to be sold to other bigger companies. As the owners were entrepreneurs and committed to the company's success, they decided to maintain the growth rate and to look up for new opportunities. The main assets were a consolidated brand reputation and distribution channels that could be explored to become even bigger. The difficult was the necessity to invest in the development of new products, including R&D and infrastructure to manufacture. The use of international suppliers could reduce the time to introduce new products into the market. The company could not sell the same products those others competitors could find abroad. With this strategy, Company A started to import products and use reverse engineering at the company's site in order to develop improvements. From this process, the Company identified parts that could be changed for items they have already certified global suppliers. Company A started requesting their direct suppliers to use a component from a supplier identified and certified by the Company A. Part of these components were developed specifically to the company. This strategy started to be implemented in 2009 and the biggest challenge was related with cultural aspect. Most of the suppliers are from Asia and they had difficult to comply with the mix of quality and cost that the Company requires.

The inclusion of new products from Chinese suppliers into the company's product portfolio can represent a risk for the company, as these may not be well accepted by the Company's customers. In order to reduce the risks of customers' rejection to these products, it developed a second brand, which is used to introduce products into the market. As these products are accepted, they start selling under the Company's main brand.

A new step on this strategy is being implemented while the data were being collected for this research. In 2011, the Company receives the first batch of a product 100% developed by the Company, including design, molding and injection tooling and produced by an Asian supplier. The company considered that this is a result of the adoption of inter-firm cooperation, not just competition. The company created entry barriers selling quality products and developing distribution channels that allowed it to adopt risk strategies. The transfer of knowledge (technical and marketing information), as in this case, is considered as a very risky operation. To avoid these risks, the company establishes agreements with the suppliers in order to assure that if they sell the products to the Company's competitors, they will sell for a bigger price. The other strategy is to fragment the value added part of the product to be manufacture in an different supplier, as when a supplier that introduce the software in an component in one country and ship them complete to be assembled in another supplier.

Case Analysis

The adoption of GS by Company A is being motivated by competitive advantages, supporting Proposition 1. According to the interviewee, "we will pull the demand (...) we became a national reference and the international market is also looking at us" and to keep competitive they need to use new technologies that can be achieved in the global supply market: "worldwide electronics – every 6 months we have a new microcontroller".

To Company A, the adoption of GS is a result of the resources available for the process. There is an effort of the top-management to support this activity as the negotiation is primarily conducted by the owners, what supports Proposition 3. This is possible as the Company is medium sized and consider that the supply strategy is central for its development/growth. The interviewee highlights that when it was a small company, it could purchase from a distributor located in Brazil, but as it grow it has to be competitive searching abroad for the original suppliers. This happens because the company's industry is global – "the evolution of the electronic area is huge – what leads us to the necessity of be aware of new technologies and suppliers".

This affirmative is also related with the supply opportunity analysis process (Proposition 5). Aspects related with supply and sourcing environment are monitored by visits to supply countries and the participation in international fairs. This process focuses on the identification of "products tendency", even though the company does not have a structured or formal interface for this effort and customer requirements.

The supplier management process already absorbed the global dimension of the supplier market and the involvement of suppliers into new products development is happening at Company A, what makes it to be especially aware with the cross-functional importance of the supplier, like the engineering department (Proposition 6). The biggest difference from local to global suppliers is the scale – the need of scale leads to source globally and the scale leads the company to be able to source from international suppliers. The use of international suppliers can be a replication of the local supplier practices based on the increase of the total quantity purchased in a search for large suppliers and the reduction of the intermediary companies.

The use of global suppliers often follow a strategy to keep 50% of the total to be purchased locally, but it is also related with the capacity of the suppliers to supply Company A's demand characteristics, such as customized products (Proposition 7). This situation is also related with the characteristics of the inputs that the Company source abroad. The company developed a model of product portfolio, but according to it, all components and products that are source globally became strategic (Proposition 8).

The necessity of qualified personnel with continuously training is a very important aspect to the adoption of GS (Proposition 9). The professionals must have knowledge of "different languages", "technical" and "negotiation skills". Knowing languages is 100% supported by the company and undergraduate and graduate courses can be 50% paid by the Company.

Related with Proposition 10, the existence of communication tools and platforms is considered important to Company A but it is highlighted that Internet is still the most important tool to search new suppliers: "Give me three days and I'll find the supplier of my competitor". These tools for information management can also be used inside the company. According to the interviewee, "when you are developing a product there is the specification, the detailed scope and the experience with the things that did not work", what is maintained in a data based directly related with the engineering area.

The risk management (Proposition 11) can be separated when it consider products that have local suppliers and those that do not have it. According to the interviewee, if there are local suppliers, the risks are the same for all competitors as "the internal competition stays at the same level for all". The inexistence of quota or others barriers that could limit the imports are seen as favorable aspects.

The analysis of external barriers (Proposition 12) is conducted by Company A based in two internal questions: (1) is it possible to produce at home country and be competitive even with the barrier, and (2) how long will that barrier possible persist? Based on these answers, the managers take the decisions.

The results achieved by Company A (Proposition 13) are related with a learning process that lead the company to start developing global operations based on the division of the value chain operations in different global locations. The knowledge about the suppliers' countries is one of the reasons that make Company 1 prepared to start these operations, followed by the experience with new suppliers sourcing process, and the flexibility and agility that the company had to develop in order to negotiate with these global companies.

Based on the analysis of the case, we can conclude that the framework allowed us to understand how GS process is being conduct by Company A. Looking at the dimensions analysis made by the investigation of the propositions of this study; it is possible to identify the key success factors for each dimension, as presented in Table 1.

Dimension	Proposition	Key-success factors
Antecedents to GS / Strategic orientation	P1: The adoption of GS is motivated by competitive advantages.	• Motivation to become a global player • Faster access to new technologies
Antecedents to GS / Organizational Structure	P2: The centralization of GS activities is related with the potential synergy between units, and their supply needs.	• Not identified at the cases
	P3: The adoption of GS implies that the availability of resources for establishing and managing the activity is relative to the organization's size and, the importance attributed to GS, including the top management support, and the industry features.	• Firm's functional structure • Global industry • Top-management support
	P4: The presence of the purchase company in the supply country motivates the adoption of GS.	• Not identified at the case

Opportunities	P5: The supply opportunity analysis process includes the investigation of the supplier company, the inputs, and the supply and sourcing environments, as well as customer requirements improving the adopting of GS.	• Visits to suppliers countries and companies • Identification of "product tendency"
Process of GS / Supplier's management	P6: The adoption of GS is related with the involvement of suppliers into new product development as the units of the company and units areas are more related, but on the other hand work with global suppliers is more difficult comparing with local suppliers.	• Establishment of interfaces with suppliers specially with the engineering area
	P7: The choice to use foreign suppliers based on the purchaser country will be related with support, customer service, and cultural aspects related to the supplier management process.	• Not identified at the cases
Process of GS / Purchase Process	P8: The GS activity is oriented to inputs consider strategic for the company, and the other inputs should be included in the context of this strategy.	• Classification of all imported inputs as strategic
	P9: The GS approach requires qualified personnel and continuously training of them in order to identify better opportunities and conduct efficient sourcing process.	• Cultural (languages), technical and negotiation skills
	P10: The existence of communication tools and platforms are important to global supply management, including actual and potential suppliers, and actual and potential demands.	• Use of internet
Difficulties and risks	P11: GS includes the management of risks, considering cultural and governmental diversities and its special characteristics.	• Development of alternative sourcing opportunities
	P12: The analysis of external barriers is important during the process of purchase opportunity analysis to avoid risks and ensure benefits.	• Trade-off analysis
Results	P13: The adoption of GS leads to competitive advantage comparing with firms that purchase internationally	• Faster product development • Introduction of new technologies

Figure 2 – Case Analysis Summary

9. DISCUSSION

The investigation of GS adoption by emerging countries companies is an opportunity to extend the knowledge developed about GS based on companies from developed countries. The use of the framework based on the previously theory contributes to carry out one single case study to illustrate the key success factors on each dimension.

Some of the propositions of the theoretical framework were not found during the case investigation. Proposition 2 was not found because it refers to the centralization of GS activities considering a company that has more than one purchase unit and Company A has just one site and one purchase area. The same problem happened with the analysis of Proposition 4 that is related with the presence of the purchase company in the supply country. While these propositions are related with the structure of Company A, the Proposition 7 was also not found. The interviewee does not highlighted aspects related with the use of foreign suppliers based on their country. The impossibility to identify results related to these propositions may occurred because of the size of the company and the centralization of operations. The characteristics of electro-electronics industry may also be a reason for it, it has global products made of global inputs that are manufactured primarily overseas (especially in Asian countries).

An aspect that was identified at the study is the capability of the company to analyses its value chain and defines to focus on the activities that add more value: product development and distribution channels. The capability to manage a global operation is being developed by Company A as an incremental process that started with the movement from local to global suppliers and it is getting intense with the total manufacture of finished products abroad. The concern with the creation of a learning process as the company goes forward on this process seems to be an important aspect of the adoption of GS by the company. The managers do have a strategic view of their global market and are trying to build competitiveness with the integration of global operations maintaining the core competence protected in order to be able to compete. The ability to develop relationships with global suppliers is also an important characteristic of Company A as it allowed them to develop a negotiation not based just on cost, but also on trust and mutual benefits.

It is important to make some considerations about the industry. The electrical and electronic industry in

Brazil is an import-based industry what can directly impact the sourcing strategy of its companies. This industry sector was chosen as it is dependent of international suppliers as their current market and their supplier market act as global even inside the Brazilian territory. This situation would lead the companies of this industry to have the possibility to choose raw materials from representative inside Brazil or purchase them abroad. This possibility of choice can make the strategy definition a complex process and lead to different approaches of sourcing.

10. CONCLUSIONS

The first contribution of the case is a framework to guide further research and the generation of insights about the adoption of GS by a Brazilian company that may represent some directions about how these companies are adopting this strategy. As this case represents the first stage of a large research project and more cases will be analyze, it is a preliminary empirical data to guide further research, even at the same industry.

Looking at the internationalization process of the investigated company, it was identified that it is moving from being just a manufacturer, to become a distributor in the Brazilian market. The production outsourcing process is a change identified in some companies in several industry sectors, like footwear, that is not as intense in technology as the electric and electronic industry (Paiva and Vieira, 2011). The presence of this change in an industry based on technology with the maintenance of the product development, or part of it, is something new in internationalization studies. This issue is a potential gap identified in the empirical research.

Company A is motivated to adopt GS by the possibility to innovate (faster access to technology) and to become a more competitive player in the Brazilian market through the control of distribution channels. At the same time, it was a way to compete with global players. These aspects were not found in the previous literature review. The role of the company's owner into the GS process was also not found in the literature review from developed countries. In this case study, we identified that the owner played a strategy role as a real entrepreneur for the international market. The owner was responsible for the beginning of sourcing internationalization process. They lead us to reformulate the Proposition 1.

P1.1: The adoption of GS is motivates by the gain of faster access to new technologies and to become a global player.

P1.2: The adoption of GS is motivated by the company's owner and its entrepreneurial capacity for international operations.

These propositions represent the second contribution of this research. They came from empirical evidence from the case study but it is not present in the current literature. These propositions can improve the theoretical framework as they reveal new aspects about the adoption of GS by companies from emerging countries that can be verified in other case studies. The results of the analysis of a single case study are not strong enough to contribute to theory development, but provide insights that, in global sectors, companies from emerging countries must have to develop global sourcing strategy to compete in their local market (Brazil) and low development countries (other South American countries). Even if this GS strategy has been successful to the Company A so far, it does not allow it to get a competitive advantage to reach more sophisticated markets. It also shows that replicating the relationship with suppliers established with local suppliers can have some barriers related to cultural differences. Companies from developed countries have already overcome these barriers, as they are more culturally diverse. Further research based on the new two propositions can expand the understanding of companies from emerging countries and GS strategies.

REFERENCES

Agndal, H. (2006) The purchasing entry process – A study of 10 Swedish industrial small and medium-sized enterprises. *Journal of Purchasing & Supply Management.* Vol.12, pp.182-196.

Alguire, M.S.; Frear, C.R.; Metcalf, L.E. (1994) An examination of the Determinants of Global Sourcing Strategy. *Journal of Business and Industrial Marketing.* Vol. 9 No. 2, pp. 62-75.

Arnold, U. (1989) Global sourcing – An indispensable element in worldwide competition. *Management International Review.* Vol. 29 No. 4, pp. 14-28.

_____. (1999) Organization of global sourcing: ways towards an optimal degree of centralization. *European Journal of Purchasing & Supply Management.* Vol. 5, pp. 167-174.

Barrat, M.; Choi, T.Y.; Li, M. (2011) Qualitative case studies in operations management: Trends, research outcomes, and future research implications. *Journal of Operations Management.* Vol 29, pp. 329–342.

Bozarth, C.; Handfield, R.; Das, A. (1998) Stages of global sourcing strategy evolution: an exploratory study. *Journal of Operation Management.* Vol. 16 , pp. 241-255.

Butter, F.A.G.; Linse, K.A. (2008) Rethinking procurement in the era of globalization. *MIT Sloan Management Review.* Vol. 50 No. 1, pp. 75-80.

Cebi, F., Bayraktar, D. (2003) An integrated approach for supplier selection. *Logistics Information Management.* Vol. 16 No. 6, pp. 395-400.

Cho, J., Kang, J. (2001) Benefits and Challenges of Global Sourcing: Perceptions of US Apparel Retail Firms. *International Marketing Review.* Vol. 18 No. 5, pp. 542-561.

Christopher, M. (2002) *Logística e Gerenciamento da Cadeia de Suprimentos.* São Paulo: Pioneira Thomson Learning.

Christopher, M., Peck, H., Towill, D. (2006) A taxonomy for selecting global supply chain strategies. *The International Journal of Logistics Management.* Vol. 17 No. 2, pp. 277-287.

Dornier, P., Ernst, R., Fender, M., Kouvelis, P. (2000) *Logística e Operações globais*: textos e casos. São Paulo: Atlas.

Dutton, G. (2008) Ten reasons to use global sourcing. *World Trade.* Vol. 21 No. 6.

Fleury, A.; Fleury, M.T.M.; Reis, G.G. (2010) El Camino Se Hace Al Caminar: La Trajectoria de lãs Multinacionales Brasileñas. *Universia Business Review.* Primer Cuatrimestre, PP.34-56.

Gammeltof, P.; Barnard, H.; Madhock, A. (2010) Emerging multinationals, emerging theory: Macro and micro-level perspectives. *Journal of International Management,* Vol. 16 No. 2, pp. 95-101.

Gelderman, C.J.; Semeijn, J. (2006) Managing the global supply base through purchasing portfolio management. *Journal of Purchasing & Supply Management.* Vol. 12, pp. 209-217.

Ghauri, P.N.; Tarnovskaya, V.; Elg, U. (2008) Marketing driving multinationals and their global sourcing network. *International Marketing Review.* Vol. 25 No. 5, pp. 504-519.

Grieco, P. L. (1995) *Supply Management Toolbox* – How to Manage your Suppliers. West Palm Beach: PT Publications, Inc.

Harris, G. L. (2006) The essence of global sourcing. *Contract Management.* Vol. 46 No. 3, pp. 16-21.

Jin, B. (2004) Achieving an optimal global versus domestic sourcing balance under demand uncertainty. *International Journal of Operations and Production Management.* Vol. 24, No. 12, pp. 1292-1305.

Johnston, W.J.; Leach, M.P.; Liu, A.H. (1999) Theory testing using case studies in business-to-business research. *Industrial Marketing Management.* Vol. 28 No. 3, pp. 201–213.

Kamath, R.R.; Liker, J.K. (1994) A second look at Japanese product development. *Harvard Business Review.* Vol. 17 No. 6.

Knudsen, M.P.; Servais, P. (2007) Analysing internationalization configurations of SMES's : The purchaser's perspective. *Journal of Purchasing & Supply Management.* Vol. 13 No. 2, pp. 137-151.

Levy, D.L. (1995) International Sourcing and Supply Chain Stability. *Journal of International Business Studies.* Vol. 26, pp. 343-360.

Lima, J.C.S. (2004) Um Estudo sobre a Reconfiguração da Função Compras em Empresas do Setor Automotivo. Tese de Doutorado. Escola Politécnica da Universidade de São Paulo. São Paulo.

Monczka, R.M.; Trent, R.J. (1991) Global Sourcing: a development approach. *International Journal of Purchasing and Materials Management.* Vol. 27 No 2, pp. 2-8.

Mulani, N. (2008) Five "good behaviors" that are key to global sourcing. *Logistics Management.* Vol. 47 No. 6.

Paiva, E.L.; Vieira, L.M. Strategic choices and operations strategy: a multiple case study. International Journal of Services and Operations Management. Vol. 10, pp. 119-2011.

Porter, M. E. (1986) *Estratégia competitiva.* Rio de Janeiro: Campus.

Quintens, L.; Pauwels, P.; Matthyssens, P. (2006) Global purchasing strategy: conceptualization and measurement. *Industrial Marketing Management.* Vol. 35, pp. 881-891.

Salmi, A. (2006) Organizing international supplier relations: An exploratory study of Western purchasing in China. *Journal of Purchasing & Supply Management.* Vol.12, pp. 197-208.

Samli, A.C, Browning, J.M. (2003) Developing Competitive Advantage. *Journal of Global Marketing.* Vol. 16 No. 4, pp. 35-52.

Sheth, J.N.; Sharma, A. (1997) Supplier Relationships – Emerging issues and challenges. *Industrial Marketing Management.* Vol. 26, pp. 91-100.

Simchi-Levi, D.; Kaminsky, P.; Simchi-Levi, E. (2003) *Cadeia de Suprimentos: Projeto e Gestão.* Porto Alegre: Bookman.

Sirkin, H.L.; Hermeling, J.W.; Bhattacharya, A.K. (2008) Globality: challenger companies re radically redefining the competitive landscape. *Strategy & Leardership.* Vol. 36, No.6, pp. 36-41.

Sislian, E.; Satir, A. (2000) Strategic sourcing: a framework and a case study. *The Journal of Supply Chain Management.*

Smith, J.M. (1999) Item selection for global purchasing. *European Journal of Purchasing & Supply Management.* Vol. 5 No. 3-4, pp. 117-127.

Steinle, C.; Schiele, H. (2008) Limits to global sourcing? Strategic consequences of dependency on international supplier: Cluster theory, resource-based view and case studies. *Journal of Purchasing and Supply Management.* Vol. 14 No. 1, pp. 3-14.

Tolentino, P.E. (2010) Home country macroeconomic factors and outward FDI of China and India. *Journal of International Management.* Vol. 16, No. 2, pp. 102-120.

Trautmann, G.; Bals, L.; Hartmann, E. (2009) Global sourcing in integrated network structures: The case of hybrid purchasing organizations. *Journal of International Management.* Vol. 15 No. 2, pp. 194-208.

Trent, R.J.; Monczka, R.M. (1998) Purchasing and supply management: trends and changes throughout the 1990s. *International Journal of Purchasing and Materials Management.* Vol. 34 No. 4,pp. 2-11.

_____. (2003) International purchasing and global sourcing – what are the differences? *Journal of Supply Chain Management.* Vol. 39 No. 4, pp. 26-36.

_____. (2003a) Understanding integrated global sourcing. *International Journal of Physical Distribution & Logistics Management.* Vol. 33 No. 7, pp. 607-629.

Trunick, P.A. (2006) Global sourcing markets, global headaches. *Logistics Today.*

Yin, R. K. (2001) *Estudo de casos: Planejamento e Métodos.* Porto Alegre: Bookman.

Zeng, A.Z., Rossetti, C. (2003) Developing a Framework for Evaluating the Logistics Costs in Global Sourcing Process. *International Journal of Physical Distribution and Logistics Management.* Vol. 33 No. 9, pp. 785-803.

Wilding, R.; Braithwaite, A. (2007) Global transactions: managing risks in global sourcing. *Supply Chain Europe.* Vol. 16, No. 2, pp. 26-29.

AUTHOR'S BIOGRAPHY

Moema Pereira Nunes is a Professor at PUCRS Business School. She holds a PhD degree in Management from Universidade do Vale do Rio dos Sinos (UNISINOS. Moema was visiting Ph.D. student at Temple University (USA). Her research interest are international business and global sourcing.

Luciana Vieira is a Professor at Unisinos Business School, Brazil. She holds a PhD degree in Agricultural and Food Economics from the University of Reading, UK. Luciana was Visiting Fellow at Brown University (USA) and at the Institute of Development Studies (UK). She has published in several international journals and her research interests are global supply chain and buyer supplier relationships.

José Antônio Valle Antunes Júnior is a Professor at Unisinos Business School and Director of Produttare Consultores Associados. He holds a BA in Mechanical Engineering from Universidade Federal do Rio Grande do Sul (UFRGS), MSc in Production Engineering from Universidade Federal de Santa Catarina (UFSC) and a PhD degree in Management from the Universidade Federal do Rio Grande do Sul (UFRGS).

3

Selecting Maritime Disaster Response Capabilities

Aruna Apte
Naval Postgraduate School
auapte@nps.edu

Keenan D. Yoho
Naval Postgraduate School
kdyoho@nps.edu

Cullen M. Greenfield
United States Navy
greenfc@lsd48.navy.mil

Cameron A. Ingram
United States Navy
cam_ingram22@yahoo.com

ABSTRACT: Using a structured, qualitative evaluation schema complemented by expert rating, we evaluate the capabilities and utility of ships in the United States Navy and United States Military Sealift inventory. We find that there are specific types of vessels with significant disaster response utility and recommend a flotilla type that would be best suited for these types of operations. Utilizing an exploratory framework that evaluates three diverse disaster cases, we scale the utility of each vessel using subject matter experts. This work should be of interest to national policy-makers as well as international governing bodies and leaders of naval institutions for its recommendations on the type of ships most useful for contributing to effective disaster response. The capabilities identified in this research are found in various naval and maritime organizations throughout the world, and this work can help assist those organizations identify the types of capabilities that are most useful in the event of a disaster.

Keywords: *Disaster Response, Capability, United States Navy, Humanitarian Operations*

1. INTRODUCTION

In the last decade, incidences of major disasters have increased significantly. In 2009, there were 335 natural disasters reported worldwide that killed 10,655 persons, affected more than 119 million others, and caused over $41.3 billion in economic damages (Vos *et al.*, 2009). Figure 1 shows the overall trend of disasters and damage by year. While the trend in the number of disasters reported shows an increase, it is not clear that there has been a commensurate response.

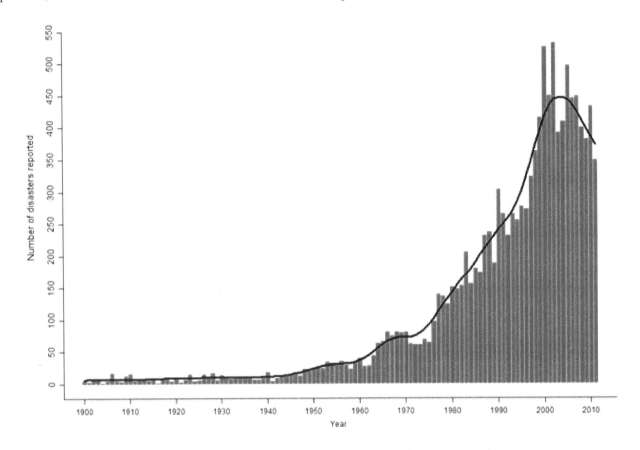

Figure 1: Natural Disasters Reported 1900–2011(EM-DAT, 2011)

Whereas matching supply to demand is an important aspect of supply chain management, disasters invariably result in conditions where both supply and demand can be extremely uncertain (Whybark, 2007) and matching is made very difficult. Webster's Dictionary defines capability as an ability to do something, the quality or state of being capable or an ability to perform actions. We, in this research, find and analyze those capabilities that lead to effective (in terms of reaching most of the affected population) and efficient (in terms of time and money) relief operations in case of disaster. We believe that understanding various organizations' abilities to match supply to demand for humanitarian operations will enrich the planning process for such relief operations, leading to efficient management of a response supply chain (Balcik *et al.*, 2010).

The United States (U.S.) military, United Nations (UN), Red Cross, CARE, World Food Program, North Atlantic Treaty Organization (NATO), and Military and Civil Defense Assets (MCDA) are but a few of the significant players in a disaster response supply chain. In this study, we examine the unique capabilities and assets of the U.S. Navy (USN) and Military Sealift Command (MSC) to determine which maritime military assets are best suited for disaster response.

1.1 Motivation

On November 15, 2007, Bangledesh was struck by Cyclone Sidr, which caused over 10,000 deaths and more than US$450 million in damages. In response, the USN decided to send one of its guided missile destroyers (DDGs), which was in the vicinity. Due to the gradual slope of the ocean floor along the Bangledesh coastline, which creates a relatively shallow draft, the DDG could approach no closer than 25 miles from the shoreline, which was out of the visible range. Additionally, the DDG was not outfitted with a helicopter, could not produce enough water to supply victims or hospitals ashore, and did not have extra food or medical supplies to provide the disaster victims. In short, this ship was unable to provide tangible relief to the devastated area, an example of a misplaced asset leading to underutilized capability.

Based on United States Agency for International Development (USAID) fact sheets and Center for Naval Analysis, U. S. forces have been diverted from original mission 366 times for humanitarian assistance as opposed to 22 times for combat from 1979 to 2000. The USN may send assets that are able to reach an affected area quickly, but the effort serves no purpose if the vessels sent are not suitable for the task at hand. Such examples illustrate that in order to become more effective in humanitarian assistance and disaster relief (HADR) operations we need to understand which assets may contribute the greatest benefit when responding to specific disasters. This was our motivation for this research since the prevailing assumption that the closest asset is the best asset to deploy will not necessarily lead to a suitable or effective utilization of resources. It is evident that although the USN has contributed many ships to past disaster relief efforts, not all are equally capable of providing effective relief nor should all be sent to respond. It is not a question of providing the relief but what capabilities exist to perform such humanitarian operations is of consequence in order to achieve efficient and effective relief. Our research objective is an effort to understand the USN's capability to meet the perceived demand of an affected population during a disaster and analyze the capabilities of all major assets in the USN and MSC inventory.

1.2 Contribution of Our Research

We studied the past responses of the USN not only to understand the capabilities but also to analyze whether these capabilities have been utilized to the best of their ability. The response to the earthquake in Haiti, Hurricane Katrina, and the Indian Ocean tsunami are but three among many contributions made by the USN to natural disasters over the past decade. With a large number of the world's population living in close proximity to the coastline, the USN's ability to traverse and deploy large quantities of relief from the sea is a valuable capability during a disaster relief operation and helps to reduce the gap of pain, which has been described as the time between the demand for aid and the time in which the aid is provided (Cuculo, 2006).

As defense budgets for some nations, to include the U.S., become constrained, it is increasingly essential that ships, aircraft, and land-based systems are adaptable for many uses, to include those that do not directly involve warfare. Current literature in the public sector (Xinhua, 2013; Miles, 2011; Guilloux, 2011; Moroney et al. 2013) illustrates the need for understanding the capabilities of organizations involved in disaster response. However, to our effort we could not identify any scholarly articles on the specific topic discussing the capabilities of organizations involved in humanitarian assistance and disaster relief. We believe, this is a new but critical topic in the field of humanitarian logistics and therefore our analysis is of value to decision makers who must determine capital equipment acquisitions that will shape naval forces that must deploy across a wide spectrum of crises. This is of special note since the U.S. Department of Defense has articulated a broad strategic vision to influence foreign policy and achieve defense goals through operations other than war, in particular, by employing HADR. We propose an "HADR flotilla" that can be better utilized for HADR operations as a step in the right direction. There has been research that looks at the USN's role in humanitarian assistance (Freedberg, 2012); however, ours is the first study of its kind that links the strategic considerations of determining what types of missions will be included as part of a security strategy (such as those that include HADR operations) and then connecting those strategic decisions to the operational options available as a result of asset purchases that took place in the past. Further, our work informs future asset purchase decisions with respect to which vessel types serve multiple uses (warfare and non-warfare missions) that contribute to national security.

2. UNDERSTANDING THE SUPPLY AND DEMAND OF A NATURAL DISASTER

In order to provide appropriate humanitarian support, an organization must assess the needs of the affected population, study its own capabilities to provide relief, and match the needs of the affected population with their own native capabilities. From the operations perspective, there have been many scholarly articles written that describe the activities that occur during disaster relief operations (e.g., Apte, 2009; Yi and Ozdamar, 2007; Hale and Moberg, 2005; Pettit and Beresford, 2005; Barbarosoglu *et al.*, 2002; Baker *et al.*, 2002; Day *et al.*, 2012.

In many disasters, supply does not necessarily follow the demand. What is needed and precisely when and where it is needed may be unknown. Further, the capabilities and scope of the supply of disaster responders may be unknown to those in the affected disaster area. The uncertainty associated with both the demand for and supply of disaster assistance frequently leads to a mismatch of goods and/or services in quantity, type, or both (Apte, 2009; McCoy, 2008).

To better understand the nature of demand and supply for aid, we analyzed different types of disasters to uncover and describe a generalized set of needs (or demand), which we call disaster traits. Some of these needs will be met by non-military organizations, and some by military organizations. On the supply side, our focus is on the military organization. We specifically study the role of the USN in the past three disasters to understand how demand was met—not through just one asset, such as a ship, but through a portfolio of assets that constitute the force structure of the USN. We also wished to understand the timing of demand and supply for disaster aid. Through our analysis of the USN response to three major disaster events, we were able to determine the "on-scene" arrival of each ship and build an illustration of the cumulative supply of assistance over time.

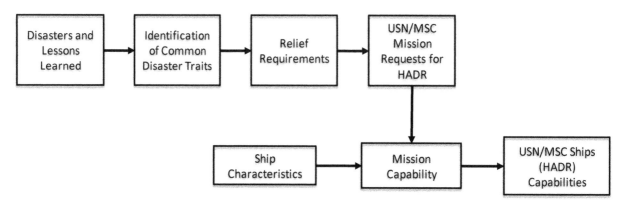

Figure 2a. Flow Chart of Demand and Supply: USN

Figure 2a shows that every disaster has certain traits that generate USN and MSC mission requests for ships, leading to a requirement for certain mission capabilities of those ships. All ships have specific characteristics that make them suitable for certain types of requirements. Once the requirements are known, then these demands—as well as the supply of ship characteristics that can meet those demands—can be matched.

Disasters and war share several traits, such as a high number of deaths and injuries, population dispersion, facility destruction and loss of common goods (such as a freshwater supply), and a need for medical facilities and personnel. Both the U.S. military

and non-military organizations bring assets, skills, and capabilities to a humanitarian crisis. The competencies and capacities of each are far from homogeneous. Identification of the specific competencies and capabilities that are core to the types of organizations will clarify who, what, and when relief is brought to the crisis. In this study, we perform the analysis for one such organization. Though the specific methodology is domain-specific in details we describe an overall process similar to ours of evaluating the capabilities of resources of the corresponding organization in Figure 2b. Such process analysis will yield what the capabilities and competencies of that organization are and whether they are utilized efficiently.

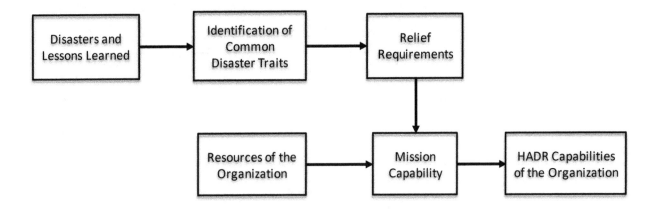

Figure 2b. Overall Flow Chart of Demand and Supply

The remainder of the paper is structured as follows. In Section 2, we describe the demand for resources during a disaster by relating a set of common disaster traits that drive relief requirements. In Section 3, through our data collection, we describe the capabilities of specific USN vessels and the mission requests that each is able to fulfill. In Section 4, we observe and identify those vessels that are most capable of fulfilling specific demands and requirements during a disaster. The fulfillment process has two dimensions: (1) the supply of a needed resource for the demand and (2) the timing of the fulfillment of that demand. In Section 5, we offer our analysis about how demand for disaster relief may best be met by specific types of USN assets based upon their capabilities and limitations. In the last section, we conclude the article and describe possibilities for future research.

3. THE DEMAND

We collected and analyzed data on the scale and scope of the following disasters: the 2004 tsunami in the Indian Ocean, the 2005 Hurricane Katrina, and the 2010 earthquake in Haiti. The primary reasons we selected these three disasters was because (1) the USN provided significant assistance and relief in each instance, (2) the disasters offered enough similarities and differences to make interesting contrasts from a case perspective, and (3) all three were ma-

jor disasters that occurred within the last decade—therefore, data were readily available.

Table 1 shows the similarities and differences between each disaster, which resulted in different demand signals for search and rescue, transfer of affected members of the population, triage facilities, and critical supplies such as freshwater. For each of the three events discussed in this research, we first developed a list of common disaster traits.

3.1 Disaster Traits

The Indian Ocean tsunami of 2004 was the result of an earthquake that measured 9.1 in magnitude on the Richter scale; at the time, it was the fourth-largest earthquake worldwide since 1900. The tsunami that occurred in the aftermath struck more than a dozen countries. Indonesia experienced most of the damage, resulting in broad destruction of many elements of critical infrastructure. The coastal highway on the island of Sumatra was completely destroyed, making many damaged areas inaccessible to land-based aid workers (Elleman, 2007). In Indonesia, the tsunami destroyed more than 25% of Aceh Province's villages, along with much of the transportation and communications infrastructure. Other countries affected by the tsunami lost all electric power production capability. Debris along the coast of Indonesia as a result of the tsunami presented navigation challenges to ships because of uncertainty in the location of the coastline.

Table 1. Effects of the Indian Ocean Tsunami, Hurricane Katrina, and the Haiti Earthquake

Disaster	2004 Indian Ocean Tsunami	2005 Hurricane Katrina	2010 Haiti Earthquake
Affected Area			
Location	Coastal	Coastal	Coastal
Scope	Dispersed in many countries	Localized in one country	Localized in one country
Infrastructure	Inadequate	More than adequate	Substandard or non-existent
Economic Development	Underdeveloped or Developing	Developed	Underdeveloped
Speed of Onset	Sudden and then Imminent	Imminent	Sudden
Consequences			
Deaths	>227,000	>1,700	92,000-220,000 estimated
Injured	>500,000	>2,000 in New Orleans alone	250,000
Missing	>2,000,000 (summary of initial reports)	>12,000 reported	20,000
Displaced	>1,500,000	>1,000,000 in gulf coast states	1,100,000
USN Involvement			
Number of Ships Deployed by USN	29	34	31
Days of Assistance	81	42	72

(VanRooyen and Leaning, 2005; CRS, 2005; Louisiana Department of Health and Hospitals, 2006; NOAA Public Affairs, 2007; Plyer, 2010; Wooldridge, 2010)

Hurricane Katrina struck Louisiana's coastline on August 29, 2005. Hurricane Katrina is recognized as the most costly hurricane ever to strike the United States, with an estimated $135 billion in damage to the gulf coast states (Plyer, 2010). As a result of Hurricane Katrina, more than three million people were left without power and thousands without freshwater as a result of broken water mains. More than 80% of New Orleans was underwater on August 31, 2005, to include the area's international airport. Highways throughout the Gulf of Mexico region were closed due to flooding, and there was considerable damage to the oil and fishing industries as a result of lost boats and rigs blown from their fixed locations.

The earthquake that impacted Haiti on January 12, 2010, was 7.0 in magnitude and lasted 35 seconds. The earthquake caused significant damage to Haiti's infrastructure and resulted in many deaths and serious injuries (Aymat, 2010). The Haiti earthquake destroyed all five medical facilities around Port-au-Prince as well as portions of the international airport. Most of the Port-au-Prince communication infrastructure and roadways as well as the seaport were destroyed during the immediate rescue operations. Debris from fallen buildings presented a significant obstacle to rescue teams attempting to reach those still trapped beneath the rubble in the immediate aftermath. The collapse of port docks and cranes in Haiti presented navigation challenges to ships because of uncertainty in the location of the coastline.

We found that all three disasters shared common traits and that these traits were also observed in many disasters catalogued by the International Disaster Database (EM-DAT; EM-DAT, 2011). The devastating effects of these disasters and the lessons learned from them clue us in on the possible needs of the affected community. The destruction of critical infrastructure, public goods (such as water), and critical facilities (such as hospitals) was apparent in each of the disasters. Some of the basic traits that these disasters have in common include a high number of deaths and injuries, population dispersion, homelessness, and a high number of missing persons, facility destruction and loss of common goods such as a freshwater supply, a need for medical facilities and personnel, and volunteers. These disaster traits induce the demand for relief. We list these in the first column of Figure 3. This identification is the first step in understanding the needs or demands of the affected community in an attempt to match the supply of relief with this demand (Figure 2a).

3.2 Relief Requirements

The disaster traits identified lead to specific relief requirements. Although not exhaustive, we identify those relief requirements that correspond to the disaster traits in the second column of Figure 3. In this figure, we match a specific disaster trait with various relief requirements.

Disaster Traits   **Relief Requirements**

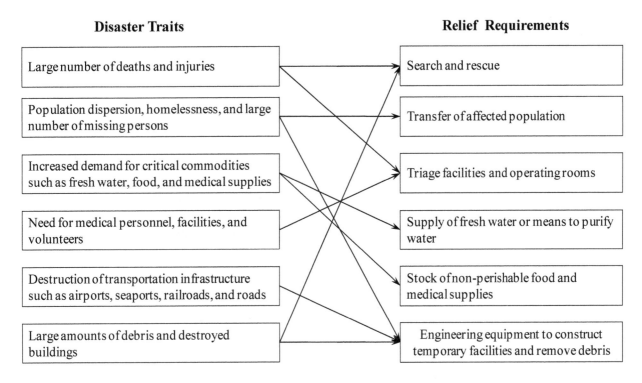

Figure 3. Basic Disaster Traits and Relief Requirements

For example, search and rescue missions are initiated as a result of missing persons; therefore, we identify search and rescue as a critical relief requirement. Homelessness and population dispersion require personnel to be transferred from unstable to stable locations. Injuries, combined with a shortage of medical supplies and clinicians, generate a requirement for trauma teams supported by triage tents and operating rooms. Damage to infrastructure such as water pipes often results in freshwater scarcity. The destruction of critical facilities and transportation infrastructure such as airports and public administration buildings generates a need for the construction of temporary structures to carry out the essential humanitarian operations. Because debris can hamper the speed and effectiveness of the relief effort, there is often a requirement for engineering equipment such as heavy earthmoving machinery. In addition to these relief requirements, there is a need for a strong and able workforce unaffected by the outcome of the disaster to conduct cleanup operations.

4. THE SUPPLY

We have discussed the needs in case of a disaster. We now describe the USN's capabilities—based on its past—in fulfilling some of these needs on the supply side. The USN has two basic classes of ships available to conduct missions: combatant ships and noncombatant ships. Combatant ships are part of the USN, and noncombatant ships are part of the MSC. The MSC has a rather unique reporting structure in its chain of command in that it reports to the commander of the USN Fleet Forces Command for Navy-specific missions, to the commander of the U.S. Transportation Command for the movement of Department of Defense items, and to the assistant secretary of the Navy for research, development and acquisition for anything related to procurement. In order to develop a comprehensive understanding of the capabilities of USN and MSC ships, we examined source documents on ships from multiple USN and MSC publications as well as the historical records of disasters and the response to those disasters by both organizations. Data supporting our analysis of the USN response to the three disasters was obtained from operational commander historical logs and archives, operational chronological records, operational orders, deployment orders, and daily situation reports. The result of our analysis is a detailed understanding of the specific characteristics attributable to each vessel in the USN and MSC inventory as well as the relative degree to which these characteristics contribute to effective disaster response.

4.1 Mission Requests

The measure of a ship's capability to conduct a specific HADR mission set is derived from the vessel's specific characteristics. The USN and MSC have many different types of vessels with different capabilities that are selected through a mission request process generated by the officers of the Navy. In order to understand which ships are better suited for HADR operations, we first discuss the different types of missions that operational commanders may be requested to conduct. Mission requests may include aircraft support, amphibious landing, or freshwater production, as well as the ability to conduct towing or salvage operations.

The specific characteristics of all USN and MSC vessels were drawn from public databases such as *Jane's Fighting Ships* (*Jane's*, 2010), the MSC handbook (MSC, 2010), and the online Navy Fact Files (www.navy.mil/navydata/fact.asp). These characteristics include speed, draft, lift capacity, number of onboard personnel, freshwater-making capacity, storage space, and other traits that enable a ship to support one of the identified HADR missions

4.2 USN and MSC Ship Capabilities

In this subsection, each ship class and its HADR-related characteristics are broken down by platform, to include the various classes within the platform. The characteristics analyzed include only those characteristics applicable to HADR operations.

Not all of the MSC's vessels are government owned nor are they all in a ready status. The MSC is capable of leasing commercial ships and maintains its own forces as well. A general breakdown of MSC vessel types includes the following: naval fleet auxiliary forces, which supply USN ships at sea; special mission ships, which perform a variety of missions; ships that enable strategic placement of military supplies in key areas of the ocean (defined as prepositioning); and sealift ships, which provide ocean transportation for equipment and supplies (MSC, 2010).

For classification purposes, we have divided MSC ships according to their program designation. The most beneficial aspects of the MSC fleet are the ability to carry large amounts of cargo to a disaster area and the ability to provide significant medical support. Another benefit of MSC ships relevant to HADR missions is that some of them are self-sustaining, in terms of their capability of on-loading and off-loading cargo without the assistance of outside equipment.

Landing craft serve as the waterborne transportation link between amphibious platforms and the shore. In HADR operations, landing craft play the critical role of getting supplies, cargo, and personnel to and from the shoreline and supporting ships. USN landing craft were not designed for HADR operations, yet they are valuable assets that are capable of supporting the mission because of their lift capacity, draft, speed, and range.

When considering specific vessels, a key facet of their capability is the seaborne aircraft. Seaborne aircraft include all helicopters and fixed-wing aircraft that may be utilized by USN and MSC ships. Among the fixed-wing aircraft the USN has in its fleet, only the MV-22 Osprey was considered for this study. It is important to note that most military fixed-wing aircraft that play any significant role in HADR operations are characteristically too large to land onboard any USN or MSC vessel and therefore play no role in determining the usefulness of different vessels. Aircraft were evaluated in terms of the capabilities they contributed to their sea-basing platform; these capabilities were primarily lift capability, personnel transportation capability, and range.

We compared the capabilities of USN and MSC platforms to basic mission requests to identify which USN and MSC vessels are best suited to satisfy the relief requirements. We evaluated the relative utility of each vessel type using ordinally scaled ratings, which are assigned by USN surface warfare officers. These ratings are based on similar principles as the color-coded operational readiness of a ship (CNIC, 2012). Our ordinal ratings have three values (instead of color coding for understanding) indicating whether a ship has "little to no capability" (○), "some capability" (◑), or "significant capability" (●) to accomplish a specific mission request. The three rating values for each of the capabilities for HADR mission requests and their operational definitions are given in Table 2.

Table 2. Capability Parameter Definitions for HADR Missions 4.3 *USN and MSC Ship Platforms*

		Capability Rating Defnition
Aircraft support	○ 0	No embarked helo; unable to support helicopter operations
	◑ 1	Single helo embarked; able to support the majority of helo platforms
	● 2	Multiple helos embarked; able to sustain multiple flight operations simultaneously
Landing Craft support	○ 0	No ability to support landing craft
	◑ 1	Some ability to support landing craft
	● 2	Landing craft embarked; able to load / off load cargo and store amphibious vehicles
Search and Rescue (SAR)	○ 0	No embarked helo; unable to efficiently conduct SAR missions
	◑ 1	Single embarked helo with communication equipment and night vision
	● 2	Multiple helos embarked with communication equipment and night vision
Dry goods storage / Refrigerated goods storage / Fresh water storage / Roll On Roll Off / Fuel storage & dispensation / Self-sufficient (Cargo Capacity)	○ 0	No ability to store goods beyond current ship crew use
	◑ 1	Ability to store supplies beyond ship crew's use
	● 2	Ability to store and transfer large quantities of supplies
Personnel transfer	○ 0	No ability to support personnel transfer; slow speed vessel with deep draft
	◑ 1	Ability to support personnel transfer for 15+ personnel
	● 2	High speed, shallow draft vessel with ability to transport 30+ personnel per voyage
Fresh water production	○ 0	No ability to produce freshwater beyond shipboard usage
	◑ 1	Ability to produce and transfer >2,000 gallons per day (gpd) beyond shipboard usage
	● 2	Able to produce and transfer > 5,000 gpd beyond shipboard usage
Personnel support	○ 0	Low crew size with minimal ability to support HADR mission (< 50 personnel)
	◑ 1	Medium size crew which can support HADR mission (51 - 200 personnel)
	● 2	Large crew with ability to support HADR mission (> 200 personnel)
Berthing capacity	○ 0	Little to no excess berthing or facilities (< 30 racks)
	◑ 1	Some excess berthing and facilities (31-50 racks)
	● 2	Large number of excess berthing and facilities (> 50)
Medical support	○ 0	No ability to conduct inpatient medical treatment; no Medical officer embarked
	◑ 1	Some medical support onboard; ability to support minor medical procedures
	● 2	Medical officer embarked; ability to perform surgeries and hold several patients
Transit speed	○ 0	0-18 knots max speed
	◑ 1	19-24 knots max speed
	● 2	25 + knots max speed
Hydrographic survey	○ 0	No ability to conduct hydrographic surveys
	◑ 1	Some ability to conduct hydrographic surveys to include soundings and chart building
	● 2	Able to conduct hydrographic surveying, soundings and chart development
Salvage Ops	○ 0	No ability to conduct salvage operations
	◑ 1	Some ability to conduct lift and salvage operations in shallow waters
	● 2	Able to conduct heavy lift and deep water salvage operations
Towing	○ 0	No ability to conduct towing operations
	◑ 1	Ability to conduct emergency towing operations
	● 2	Designed to conduct push, pull, or alongside towing operations

The USN vessels (some examples are shown in Figure 4) are displayed in four categories in Table 3: (1) nuclear-powered aircraft carriers (Nimitz and Enterprise class), (2) amphibious ships, (3) cruisers and destroyers (CRUDES), and (4) other, including the littoral combat ships, patrol craft, and mine countermeasures ships. Figure 4. Amphibious Assault Ship,

Hospital Ship, and Roll-on/Roll-off Ship

For the MSC, naval fleet auxiliary force vessels are made up of fleet replenishment oilers, dry cargo and ammunition ships, rescue and salvage ships, fleet ocean tugs, and hospital ships (see Table 4). The special mission ships are composed of a command ship, submarine tenders, ocean surveillance and survey, missile range instrumentation, and cable repair. The prepositioning ships include large, medium-speed, roll-on/roll-off vessels; maritime prepositioning ships capable of loading and unloading container-ized cargo without the aid of dockside cranes; as well as accommodating roll-on/roll-off cargo such as vehicles, offshore petroleum discharge ships, break-bulk ships for non-containerized cargo, and high-speed vessels, which are high-speed catamarans ca-pable of transporting personnel or equipment. The vessels, part of the sealift program, include large,

medium-speed, roll-on/roll-off vessels; common-use tankers; dry cargo ships; and arctic tankers. The last type, Ready Reserve Force ships, include roll-on/roll-off ships; crane ships; lighterage-aboard ships, which function as floating barges and are critical for moving cargo ashore where infrastructure has been either damaged or destroyed or is non-existent; and aviation logistics support vessels.

Table 3. USN Platforms to Capability Comparison- Our evaluation of each of the USN and MSC vessel type in terms of its ability to conduct HADR missions is described in Tables 3 and 4. From these tables, we can identify those ships that have consistently high levels of capability with respect to HADR mission capabilities by identifying the rows of full circles (encircled in the tables).

Table 4. MSC Platforms to Capability Comparison

			Capabilities																		
						Cargo Capacity															
			Aircraft support	Landing Craft support	Search and Rescue	Dry goods storage	Refrigerated goods storage	Fresh water storage	Roll On Roll Off	Fuel storage & dispensation	Self Sufficient	Personnel transfer	Freshwater Production	Personnel support	Berthing capacity	Medical support	Transit speed	Hydrographic survey	Salvage Operations	Towing
Military Sealift Command (MSC) Ships	Naval Fleet Auxiliary Force	T-AOE: fast combat support ships	1	0	0	2	2	2	0	2	1	0	0	0	0	0	2	0	0	0
		T-AO: fleet replenishment oilers	1	0	0	2	2	2	0	2	1	0	0	0	0	0	1	0	0	0
		T-AE: dry cargo and ammunition ships	1	0	0	0	2	0	1	0	2	1	0	0	0	0	1	0	0	0
		T-AKE: dry cargo and ammunition ships	1	0	0	2	2	2	1	0	2	1	0	0	0	0	1	0	0	0
		T-ARS: rescue and salvage ships	0	0	0	0	0	0	0	0	2	0	0	0	0	0	0	0	2	2
		T-ATF: fleet ocean tugs	0	0	0	0	0	0	0	0	1	0	0	0	0	0	0	0	1	2
		T-AH: hospital ships	1	0	0	0	0	0	0	0	0	1	2	2	2	2	0	0	0	0
	Special Mission Ships	LCC: command ship	1	0	0	1	1	0	0	0	0	0	0	1	1	1	0	0	0	0
		AS: submarine tenders	0	0	0	0	0	0	0	0	1	0	0	1	0	1	0	0	0	1
		T-AGOS: ocean surveillance and survey	0	0	0	0	0	0	0	0	0	0	0	0	0	0	1	0	0	1
		T-AGS: ocean surveillance and survey	0	0	0	0	0	0	0	0	0	0	0	0	0	0	2	2	2
		T-AGS: ocean surveillance and survey	0	0	0	0	0	0	0	0	0	0	0	0	0	0	0	0	0
		T-AGM: missile range instrumentation	0	0	0	0	0	0	0	0	0	0	0	0	0	0	0	0	0	0
		T-ARC: cable repair	0	0	0	0	0	0	0	0	0	0	0	0	0	0	1	0	0	1
	Prepositioning Ships	LMSR: roll-on/roll-off vessels	1	0	0	2	2	2	2	2	2	0	1	0	1	0	1	0	0	0
		MPS: roll-on/roll-off cargo such as vehicles	1	0	0	2	2	2	2	2	2	0	1	0	1	0	1	0	0	0
		MPF: loading and unloading Container cargos	1	0	0	2	2	2	2	0	2	0	1	0	1	0	1	0	0	0
		T-AOT: fleet replenishment tankers	0	0	0	0	0	0	0	2	2	0	0	0	0	0	0	0	0	0
		T-AK (USAF): container, roll-on roll-off ships	1	0	0	2	2	2	0	2	2	0	1	0	1	0	1	0	0	0
		T-AK (USA): container, roll-on roll-off ships	1	0	0	2	2	2	2	0	2	0	1	0	1	0	1	0	0	0
		T-AVB: aviation logistics support ships	1	0	0	2	2	2	0	0	2	2	0	0	1	0	1	0	0	0
		OPDS: offshore petroleum discharge ships	0	0	0	0	0	0	0	0	0	0	0	0	0	0	0	0	0	0
		Break-Bulk	1	0	0	0	1	1	1	0	0	2	0	0	0	0	0	0	0	0
		HSV: high-speed vessels	0	0	0	0	0	0	0	0	0	2	0	0	0	2	0	0	0	0
	Sealift Ships	LMSR: roll-on/roll-off vessels	1	0	0	2	2	2	2	2	2	0	1	0	1	0	1	0	0	0
		T-5: arctic tankers	0	0	0	0	0	0	0	2	2	0	0	0	0	0	0	0	0	0
		Common Use Tanker	0	0	0	0	1	1	2	0	2	1	0	0	0	0	0	0	0	0
		Dry Cargo	0	0	0	2	2	1	0	1	2	0	0	0	0	0	0	0	0	0
	Ready Reserve Force	Fast Sealift Ship	1	0	0	2	2	2	0	2	1	0	0	0	0	0	2	0	0	0
		RO/RO ships	1	0	0	2	2	2	2	2	2	0	1	0	1	0	1	0	0	0
		Crane Ships	0	0	0	2	2	0	0	1	2	0	0	0	0	0	0	0	0	0
		Lighterage-aboard ships	0	0	0	2	2	0	0	1	2	0	0	0	0	0	0	0	0	0
		OPDT: offshore petroleum discharge tankers	0	0	0	0	0	0	0	2	2	0	0	0	0	0	0	0	0	0
		Break-Bulk Ships	1	0	0	0	1	1	1	0	0	2	0	0	0	0	0	0	0	0
		Aviation Logistics Support	1	0	0	2	2	2	0	2	2	0	0	0	1	0	1	0	0	0

5. OBSERVATIONS

Our observations, based on the data collected and analyzed for the USN as well as on MSC assets, focus on timing capabilities in addition to the type of vessels deployed based on our synthesis of USN assets' suitability in HADR missions.

5.1 USN Vessels

The results of the ordinal ratings of ships to deliver HADR mission capabilities show that the USN (see Table 3) has significant cargo transfer capabilities as a result of the large numbers of helicopters and landing craft that may be deployed, the ability to bring large numbers of trained personnel to a disaster scene to assist with rescue, berthing space for temporary housing, and the large capacity to provide medical services. The amphibious assault ships have consistently high levels of capability to conduct HADR missions. Such ships may accommodate large numbers of helicopters as well as personnel and have medical and surgical capabilities aboard, to include operating rooms and a 1,000-unit blood bank. Additionally, amphibious assault ships have approximately 2,000 embarked Marines, who may be deployed to provide security and assistance to the affected area.

The limitations of cruisers, destroyers, patrol craft, and the littoral combat ship include a lack of storage and berthing facilities, insufficient medical services, lack of freshwater production or storage, and a lack of roll-on/roll-off capacity. In general, these ships are capable of traversing the oceans at high speeds but have very few of the other capabilities that are critical to HADR missions. The CRUDES—vessels that have embarked helicopters (cruisers, Flight IIA destroyers, and frigates)—may provide some aircraft support, such as search and rescue capability and personnel transfer capability. However, without an embarked helicopter, the CRUDES class of vessels is limited in their contribution to HADR missions.

5.2 MSC Vessels

The major advantage of the MSC fleet is its ability to carry large quantities of cargo to and from the disaster region (see Table 4). The MSC fleet also has two hospital ships that provide high levels of medical support as well as berthing capacity and the ability to produce freshwater. Although the MSC fleet has many desirable traits for HADR missions, it typically lacks embarked helicopters and is therefore marginally capable at conducting search and rescue missions or aircraft support. Another aspect of the MSC fleet is that the ships' crews are small, with the majority being civilian mariners. The small crews do not allow for a significant level of personnel support during HADR missions, and beyond the hospital ships, the MSC fleet does not have any capability to conduct medical support.

5.3 Capabilities and Timing

An important aspect of the understanding of the demand-supply picture is when the need arises and when it is fulfilled. In order to understand how USN HADR capabilities have been deployed in the past, as well as the types and levels of capability provided, we collected data about the on-scene arrival of every ship that was deployed to respond to the

Indian Ocean tsunami in 2004, Hurricane Katrina in 2005, and the Haiti earthquake in 2010.

Using the capability parameters given in Tables 3 and 4, we assigned each ship a capability value of 0 if it possessed little to no capability of performing a mission, a value of 1 if it had some capability, and a value of 2 if it was significantly capable of performing a mission. It is important to recognize that there is a great deal of difference between a ship with no landing craft support capability (0) and a ship with some capability (1). However, the difference between a ship with some capability (1) and a ship with significant capability (2) is much less.

We summed the capability values of all USN and MSC ships that were present on the scene during the disaster responses to arrive at a cumulative capability value for the HADR mission. The cumulative HADR capability value is a measure of the total of disaster relief capability that is on station in the disaster zone at a given time. Such measures of benefit rendered by a specific ship are in common use in the USN for certain primary events requiring specific force composition (Brown et al., 1989; Dugan, 2007; Hallmann, 2009; Silva, 2009). For instance, Brown et al. (1989) used the capability rating for the scheduling of ships for Atlantic Fleet combatants.

The cumulative capability offers an explanation of the total certain capabilities available (supply) for a given relief requirement (demand). For example, if only one ship was present during the disaster response and it had a mission capability of 1 for aircraft support, then the mission's cumulative capability for that day would be 1. If another ship were present with a mission capability score of 2 for aircraft support, then the mission's cumulative capability for that day would be 3. Such cumulative capability for aircraft support aligns with the demand for search and rescue as well as distribution of critical commodities. We used cumulative capability for each day and each mission to analyze response patterns for all three disasters that are presented in Figures 5, 6, and 7. The cumulative capabilities include critical supplies aligning with demand for fuel, dry and refrigerated goods, freshwater production, and freshwater cargo. Transportation and rescue align with the need for aircraft support that in turn facilitates search and rescue, landing craft support, and personnel transfer. Medical and shelter care capability indicate providing for berthing capacity and medical support. In addition to these cumulative capability values depicting the fulfillment of relief requirements, timing of such delivery—which is of critical importance—is illustrated along a time line in Figures 5, 6, and 7.

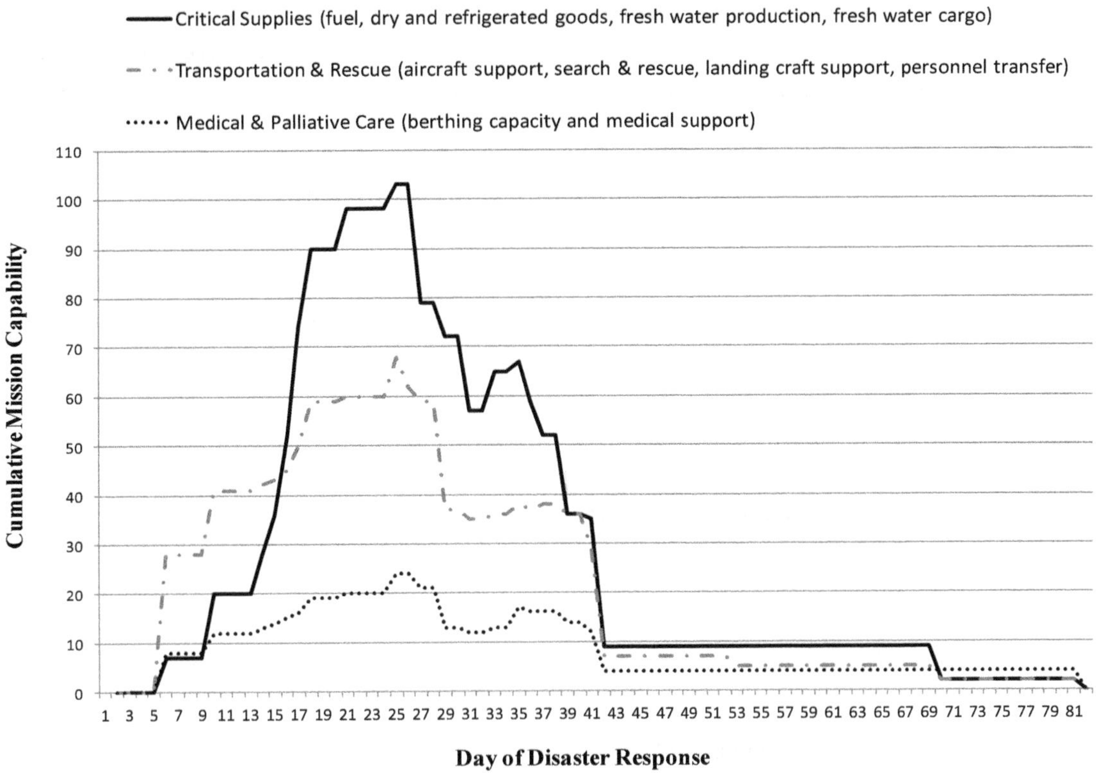
Critical Supplies (fuel, dry and refrigerated goods, fresh water production, fresh water cargo)
Transportation & Rescue (aircraft support, search & rescue, landing craft support, personnel transfer)
Medical & Palliative Care (berthing capacity and medical support)
Cumulative Mission Capability
Day of Disaster Response

Figure 5. Cumulative Mission Capability for the Indian Ocean Tsunami

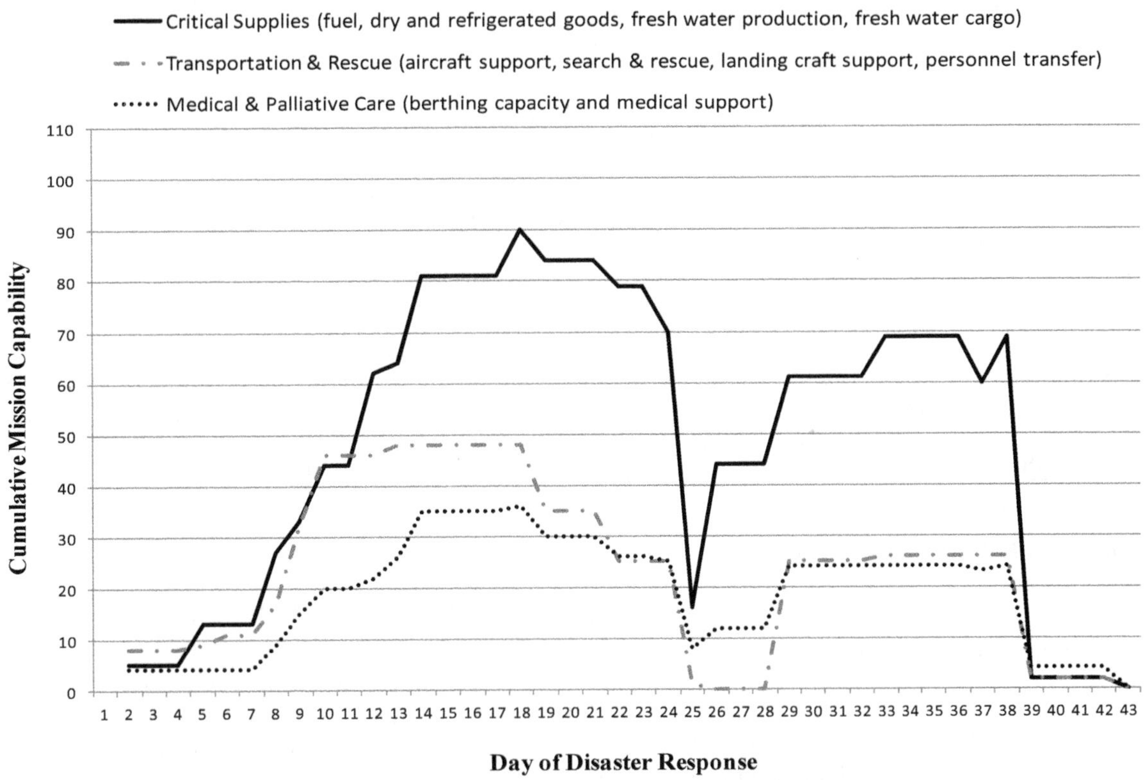
Critical Supplies (fuel, dry and refrigerated goods, fresh water production, fresh water cargo)
Transportation & Rescue (aircraft support, search & rescue, landing craft support, personnel transfer)
Medical & Palliative Care (berthing capacity and medical support)
Cumulative Mission Capability
Day of Disaster Response

Figure 6. Cumulative Mission Capability for Hurricane Katrina

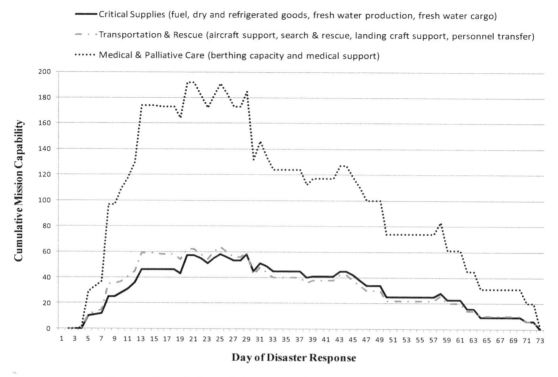

Figure 7. Cumulative Mission Capability for the Haiti Earthquake

5.3.1 Indian Ocean Tsunami

Table 5 shows that the response to the Indian Ocean tsunami lasted 81 days and that the first substantial increase in cumulative capability occurred two weeks into the effort. The peak of cumulative capability occurred on Day 24 of the response effort, and the peak range of support occurred from Day 15 to Day 35. Our data suggested that providing critical supplies was the mission that received the most support.

Table 5. Summary of the Response to Disasters

Disaster	Response		Peak of Cumulative Capability		Missions Receiving the Most Support	Missions Receiving the Least Support
	Range	Completion	Range	Occurred		
Indian Ocean Tsunami	81 Days	Day 41	Day 15 - Day 35	Day 24	Provision of Critical Supplies	Medical Care and Shelter
Hurricane Katrina	42 Days	Day 38	Day 10 - Day 23	Day 17	Provision of Critical Supplies	Medical Care and Shelter
Haiti Earthquake	72 Days	Day 41	Day 12 - Day 28	Day 19	Medical Care and Shelter	Provision of Critical Supplies

5.3.2 Hurricane Katrina

The response to Hurricane Katrina lasted 42 days, and the first substantial increase in cumulative capability occurred two weeks into the effort. All USN and MSC support for the response was complete by Day 38. The peak of cumulative capability of the response occurred on Day 17 of the response effort, and the peak range of support occurred from Day 10 to Day 23.

5.3.3 Haiti Earthquake

The USN response to the Haiti earthquake lasted 72 days. Nearly all USN and MSC support for the response effort was complete by Day 41. The first substantial increase in cumulative capability occurred two weeks into the effort. The peak cumulative mission composite capacity of the response occurred on Day 19, and the peak range of support occurred from Day 12 to Day 28.

5.4 Summary of Disaster Responses

The USN fleet is composed of carrier strike groups, expeditionary strike groups, amphibious ready groups, and submarine strike groups. These organizational structures were conceived in order to be effective in major war campaigns; however, they do not appear to be the best formations of vessels for responding to HADR operations. Our data show that the smallest number of ships used in any of the three disaster responses discussed was 29, and the largest was 34.

Studying the details of the three missions, critical supplies, transportation and rescue, and medical and shelter, we observed that in the case of the 2004 Indian Ocean tsunami, the USN deployed full expeditionary strike groups, which resulted in tasking several CRUDES vessels to support the HADR mission. CRUDES vessels provide little assistance in HADR operations and, therefore, were being underutilized for their entire time on station.

5.5 Cumulative Mission Capability

Figure 8 illustrates the cumulative mission capability from all ships present in each HADR response. The graph shows that the USN and MSC provide significantly more cumulative capacity to HADR missions within two weeks of the event and then taper off on or just before the 40th day after the disaster event. Typically, support is ramped up in the first two weeks of a response and then begins to level out as it peaks, usually near the end of the third week.

We fitted the graph of average cumulative capability over the disasters with a Weibull probability distribution. Figure 8 shows the basic shape of the relationship between cumulative capability and time in a USN and MSC disaster response effort. The slope suggests a response reaching the highest peak by Day 19 and then tapering off. This general shape of USN and MSC support to HADR missions is consistent with Pettit and Beresford's (2005) model of emergency recovery that examined emergency relief operations in military and non-military circumstances. The relationship between the "when" of the demand and the supply indicates the gap of pain due to a lag in arrival of relief. However, it should be pointed out that the bulk of the core relief arrives relatively quickly, given the size of the fulfillment, and leaves fairly quickly.

6. DISCUSSION

The USN and MSC responses to the tsunami, hurricane, and earthquake were surprisingly consistent. One hospital ship was used in each of the disaster responses. More MSC vessels were sent to each disaster than USN vessels. The number of amphibious assets employed was usually much larger than CRUDES. The one instance where more CRUDES were sent than amphibious ships was when the USN deployed an entire carrier strike group. However, since the 2004 tsunami, seemingly, the USN has adopted a joint task force approach to HADR missions. One conclusion that can be drawn from the mission response data represented in Figure 8 is that the USN and MSC typically arrive en masse two weeks into a response effort and provide peak capability during the third week. The relief rapidly declines as vessels are withdrawn during the close of an operation.

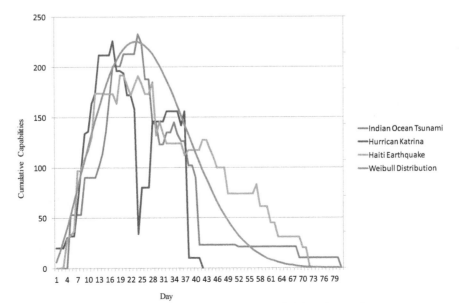

Figure 8. Cumulative HADR Capabilities for the Three Disasters

It is true that the degree of uncertainty that exists in both the disaster environment and the requirements associated with alleviating its effects greatly complicates the planning and response process. Based upon our analysis of HADR capabilities, we believe that task force composition, force structure, and design of USN and MSC HADR responses need to be robust with respect to both the environment and the requirements for assistance. Tables 4 and 5 suggest which vessels have the greatest utility when conducting HADR missions, and what combination of vessels are required to cover the full range of missions that HADR operations require. Our analysis indicates that amphibious vessels typically have the greatest utility with respect to conducting HADR operations due to their embarked helicopter capabilities, landing craft, and berthing and medical capacity. Our analysis has also shown the limitations of specific USN vessels when it comes to conducting the full range of HADR missions. In addition, the analysis validated that those vessels which can conduct lift operations are best suited to conduct HADR missions. Vessels that can provide personnel to assist in disaster response, and provide medical services, are also of great utility. However, those vessels that are limited in their ability to conduct such missions are not well suited for HADR missions.

We believe that for future HADR operations, a HADR task force composed of amphibious vessels and MSC vessels will be more effective and efficient. A task force should be able to conduct all required HADR missions with only the most effective platforms, leaving other vessels free to perform other USN missions or training. Based on Tables 4 and 5, we identified the specific vessel (classes) best suited for HADR operations that are supported by their respective capabilities. These are as follows: We believe that amphibious assault and transport dock ships need to be part of the "HADR flotilla" because of their ability for search and rescue, aircraft and landing support, freshwater production, berthing capacity, and medical support. Although nuclear-powered aircraft carriers have the ability to provide aircraft support, freshwater production, berthing capacity, and medical support, these vessels primarily support fixed-wing fighter aircraft that do not have the same level of utility as vertical lift and rotary-wing aircraft, which are abundant on amphibious assault ships. Combat ships such as destroyers and cruisers can travel at high speeds, and they have very little utility in the disaster zone because they lack the ability to produce large amounts of freshwater, do

not have excess food stores, and generally lack extra berthing capacity. These ships are particularly ill-suited to HADR missions if, in addition, they lack an embarked helicopter. The oilers, prepositioning ships with roll-on/roll-off capability, and hospital ships can also be elements of an HADR flotilla because they have storage capacity and capabilities that other commercial ships and warships lack.

The response to Hurricane Katrina provides an obvious caveat to our proposed force structure for HADR operations. The presence of nine mine countermeasures vessels in the Gulf of Mexico were capable of locating and corralling oil rigs that had broken free and were drifting in open water (United States Northern Command, 2005). Another special consideration in HADR operations—based on the 2011 Japan earthquake—is the use of nuclear-reactor–qualified personnel in gaining stability to the affected reactors, a very unique situation in which the USN played a vital role in assisting the affected population.

7. CONCLUSION

Providing appropriate HADR requires an understanding of the needs of the affected population. In this research, we proposed a list of general relief requirements (demand) based on representative disasters. We described the capabilities of the USN and MSC as suppliers of relief in disasters. Additionally, we illustrated the response time of the supply of relief. We identified the primary mission requests received by the USN and MSC and then rated each HADR mission capability provided by every ship in the USN and MSC. We used the mission capabilities for each ship and combined them with the time line of every ship deployed for each of the disasters we studied. Based upon the relief capability and response time, we calculated a cumulative capability that constitutes an index of the relative HADR mission capability for a disaster response in a particular time window.

The USN provides substantial HADR due to its unique capabilities. Our analysis in this research contributes to a firmer understanding of which vessels are most appropriate for use during disaster response. We believe our methodology may be used as a template by other military organizations that have missions to support humanitarian operations, such as NATO or the MCDA. Studies of this nature may expose the issue of relief requirements in terms of "what" and "when" for the organizations with assets similar to the USN. Such military organizations can make more effective decisions when considering the types of vessels they will procure in the future as well as in planning and executing humanitarian operations throughout the world. This is important in reducing the gap of pain since needs assessment is a critical issue, which, due to its complexity, has not been adequately studied.

Understanding capabilities facilitates the planning of humanitarian operations, especially when diverse organizations participate in the disaster relief. Such understanding facilitates the relief operations thus directly influencing the welfare of the affected population in the society. Similar evaluations of resources and the capabilities of organizations that are significant players in humanitarian operations, such as UNICEF, CARE, WFP, and Fleet-Forum, will help to identify which organizations are capable of providing what type of relief and where. Since we believe that understanding the capabilities of these organizations will develop a richer picture of humanitarian operations and effort, our research for one such organization, the USN, sets the stage for the evaluation of other organizations. This is the first step in that direction. In this vein of research other researchers can pursue similar studies to further this research by continuing the evaluation process described for disaster response capabilities.

Future research may involve the development of an optimization model to determine an appropriate "product mix" of vessels for disaster response. Such a model would determine the optimal mix of ships based on capabilities, as opposed to the proximity or availability of a ship. The model currently being developed is as follows: the objective is to reduce the gap of pain (by maximizing the available capabilities for relief requirements). The constraints result from issues such as budget/cost, availability of personnel, and operational abilities.

REFERENCES

Apte, A. 2009. "Humanitarian logistics: A new field of research and action", *Foundations and Trends in Technology, Information and Operations Management*, Vol. 3(1):1–99.

Apte, A. and Yoho, K. 2011. "Capabilities and competencies in humanitarian operations", paper presented at the 18th European Operations Management Association (EurOMA) Conference.

Aymat, J. 2010. "Operation unified response: Joint task force—Haiti, D+37 brief", unpublished PowerPoint briefing, Naval War College, Newport, RI.

Baker, S.F., Morton, D.P., Rosenthal, R.E. and Williams, L.M. 2002. "Optimizing military airlift", *Operations Research*, Vol. 50(4):582–602.

Balcik, B., Beamon, B.M., Krejci, C.C., Muramatsu, K.M. and Ramirez, M. 2010. "Coordination in humanitarian relief chains: Practices, challenges and opportunities", *International Journal of Production Economics*, Vol. 126(1):22–34.

Barbarosoglu, G., Ozdamar, L. and Cevik, A. 2002. "An interactive approach for hierarchical analysis of helicopter logistics in disaster relief operations", *European Journal of Operational Research*, Vol. 140(1):118–133.

Brown, G., Goodman, C. and Wood, K. 1989. "Annual scheduling of Atlantic Fleet naval combatants", *Operations Research*, Vol. 38(2):249–259.

Commander, Navy Installation Command (CNIC) 2012. available at: http://www.mors.org/UserFiles/file/meetings/07avi/dunn-3.pdf .

Congressional Research Service (CRS) 2005. *Hurricane Katrina: DoD disaster response*. RL33095, available at: http://www.fas.org .

Cuculo, A. 2006. "National response to catastrophic events: Applied lessons for consequence management", PowerPoint briefing presented at the meeting of Joint Center for Operational Analysis, Colorado Springs, CO.

Day, J.M., Melnyk, S., Larson, P., Davis, E. and Whybark, D.C. 2012. "Humanitarian and disaster relief supply chains: A matter of life and death", *Journal of Supply Chain Management*, Vol. 48(2):21–36.

Department of the Navy (DoN) 2010. *Naval operations concept, Implementing the maritime strategy*, available at: http://www.navy.mil/maritime/noc/NOC2010.pdf

Dugan, K. 2007. *Navy Mission Planner*, master's thesis, Naval Postgraduate School, Monterey, CA.

Elleman, B.A. 2007. *Waves of Hope: The U.S. Navy's Response to the Tsunami in Northern Indonesia*, Naval War College Press, Newport, RI.

EM-DAT 2011. "The international disaster database", available at: http://www.emdat.be/database

Freedberg, S. J. 2012. Former CNO, DepDevDef Fight to Stop Cuts to Navy's Humanitarian Mission. *Sea*. Retrieved from http://defense.aol.com/2012/06/04/hamre-roughead-fight-to-save-navys-humanitarian-mission-from-c/

Guilloux, A. 2011. "Disaster Response Mechanism: In Search of a More Effective System", *World Politics Review*, available at: http://www.worldpoliticsreview.com/articles/9588/disaster-response-mechanisms-in-search-of-a-more-effective-system

Hale, T. and Moberg, C.R. 2005. "Improving supply chain disaster preparedness: A decision process for secure site location", *International Journal of Physical Distribution & Logistics Management*, Vol. 35(3):195–207.

Hallmann, F. 2009. *Optimizing Operational and Logistical Planning in a Theater of Operations*, master's thesis, Naval Postgraduate School, Monterey, CA.

Inspector General, U.S. Department of Homeland Security 2006. *Management advisory report on the acquisition of cruise ships for Hurricane Katrina evacuees.* GC-HQ-06-11, available at: http://www.dhs.gov/xoig/assets/katovrsght/OIG_GC_HQ_06-11.pdf.

Jane's 2010. *Jane's Fighting Ships: The authority on the world's navies,* available at: http://jfs.janes.com/public/jfs/index.shtml

Louisiana Department of Health and Hospitals (2006).

McCoy, J.H. 2008. "Humanitarian response: Improving logistics to save lives", *American Journal of Disaster Medicine,* Vol. 3(5):283–293.

Miles, D. 2011. "Pacific Mission Tests Disaster Response Capabilities", *American Forces Press Services,* available at: http://www.defense.gov/News/NewsArticle.aspx?ID=62996

Military Sealift Command (MSC) 2010. *The U.S. Navy's Military Sealift Command Handbook 2010,* Author, Washington Navy Yard, Washington, DC.

Moroney, J. D. P., Pezard, S., Miller, L. E., Ergstrom, J. G., and Doll, A. 2013. "Lessons from DoD Disaster Relief Efforts in the Asia-Pacific Region", available at: http://www.rand.org/content/dam/rand/pubs/research_reports/RR100/RR146/RAND_RR146.pdf

National Oceanic Atmospheric Administration Public Affairs 2007. "Hurricane Katrina", available at: http://www.katrina.noaa.gov

Pettit, S.J. and Beresford, A.K. 2005. "Emergency relief logistics: An evaluation of military, non-military and composite response models", *International Journal of Logistics: Research and Applications,* Vol. 8(4):313–331.

Plyer, A. 2010. "Facts for features: Hurricane Katrina impact", press release, available at: http://www.gnocdc.org/Factsforfeatures/HurricaneKatrinaImpact/index.html

Schulte, A. 2010. "Henson clears path for Haiti", *Sealift,* available at: http://www.msc.navy.mil/sealift

Semilla, F. 2011. Contingency Operations Officer, Pacific Fleet, United States Navy. *Personal communication.*

Silva, R.A. 2009. *Optimizing Multi-Ship, Multi-Mission Operational Planning for the Joint Force Maritime Component Commander,* master's thesis, Naval Postgraduate School, Monterey, CA.

United States Northern Command 2005. *USNORTHCOM Hurricane Katrina Timeline—Draft,* available at: http://www.northcom.mil/FOIA

VanRooyen, M., & Leaning, J. 2005. After the tsunami—Facing the public health challenges. New England Journal of Medicine, 352, 435–438. Retrieved from http://www.nejm.org

Vos, F., Rodriguez, J., Below, R. and Guha-Sapir, D. 2009. *Annual Disaster Statistical Review 2009: The Numbers and Trends,* Centre for Research on the Epidemiology of Disasters, Brussels, Belgium.

Whybark, D.C. 2007. "Issues in managing disaster relief inventories", *International Journal of Production Economics,* Vol. 108(1-2):228–235.

Wooldridge, M. 2010. "Haiti will not die, President Rene Preval insists", available at: http://news.bbc.co.uk/2/hi/americas/8511997.stm

Yi, W. and Ozdamar, L. 2007. "A dynamic logistics coordination model for evacuation and support in disaster response activities", *European Journal of Operational Research,* Vol. 179: 1177–1193.

Xinhua 2013. "State Counselor urges improving disaster relief capabilities", *Global Times,* 2013-5-12, available at: http://www.globaltimes.cn/content/780954.shtml#.UkG7ioasim4

AUTHOR'S BIOGRAPHY

Aruna Apte, Naval Postgraduate School, is an Associate Professor at the Naval Postgraduate School in Monterey, California. She has more than 20 publications in peer-reviewed academic journals such as Interfaces, Naval Research Logistics, Production and Operations Management, a monograph in Humanitarian Logistics and one patent. Currently her research is focused in humanitarian and military logistics and she advises emergency planners in preparing for disaster response. She served as the president for Humanitarian Operations and Crisis Management College in the Production and Operations Management Society. Aruna received her Ph.D. in Operations Research from Southern Methodist University in Dallas.

Keenan D. Yoho, Naval Postgraduate School, is an Assistant Professor at the Naval Postgraduate School in Monterey, California. Dr. Yoho His primary research focus is on the allocation of resources under conditions of uncertainty and scarcity. Recent areas of focus have been on supply chain management, military special operations, and humanitarian assistance and disaster response. Keenan earned his Ph.D. in Operations Management, an M.B.A. in Operations and Information Management, and a M.S. in Industrial Relations from the University of Wisconsin-Madison as well as an undergraduate degree in Religion from Temple University.

Cullen M. Greenfield, United States Navy, is a Surface Warfare Officer in the United States Navy. Cullen has served several tours on surface warfare vessels to include Chief Engineer on the U.S.S. Tortuga. Cullen earned his Master's of Business Administration at the Naval Postgraduate School.

Cameron A. Ingram, United States Navy, is a Surface Warfare Officer (SWO) in the United States Navy. Cameron was commissioned in June of 2004 after graduating from the U.S. Merchant Marine Academy in Kings Point, NY. He has served onboard two guided missile destroyers the USS HOPPER (DDG 70) and the USS MITSCHER (DDG 57). Cameron earned his Master's of Business Administration at the Naval Postgraduate School.

4

Quantitative Modeling in Practice: Applying Optimization Techniques to a Brazilian Consumer Packaged Goods (CPG) Company Distribution Network Design (Technical Note)

Gustavo Corrêa Mirapalheta

gustavo.mirapalheta@fgv.br

Flavia Junqueira de Freitas

flaviajunqueirafreitas@gmail.com

ABSTRACT: This article aims presenting an example of quantitative modeling and optimization techniques application to the design of the distribution network of a consumer packaged goods company in São Paulo, Minas Gerais and Paraná states, Brazil. This study shows that economies of 5% to 10% (which represent in absolute terms, approximately R$10 million) can be quickly achieved by the application of linear optimization technics showing a vast area of improvement for Brazilian economy, with minimal investments, on a macroeconomic scale. First, it is made a brief review of quantitative modeling techniques as they are applied in the modeling and optimization of network problems. In the second section it is depicted the company's distribution problem.. The model is then optimized through a series of software so the methodologies and results can be compared. The article finishes with the results that the company got from the model deployment, presenting a clear case of optimization techniques in a real world application, showing the viability of easily using such techniques in a broad range of distribution and logistics problems.

Keywords: logistics, network design, optimization, quantitative modeling

1. INTRODUCTION

The usage of quantitative modeling to describe problems in the area of supply chain management is an intense research subject (McGarvey & Hannon, 2004). Several linear (and nonlinear as well) optimization techniques have been specifically tailored to them, allowing managers and researchers to have them applied in a variety of different situations (Geunes & Panos, 2005); (Winston, 2003). The usage of these techniques by a broad, non-technical audience have been much increased through the dissemination of spreadsheet software, like Microsoft Excel, and Excel's Add-In package Solver from Frontline Systems (Ragsdale, 2008). Full scale, industrial models have been studied and solved in microcomputers, through the usage of numerical simulation and optimization software like Mathworks Matlab (Radhakrishnan, Prasad, & Gopalan, 2009), (Huang, 2012), (Huang & Kao, 2012), (Eshlaghy & Razavi, 2011), Opti Optimization Toolbox (Wilson, Young, Currie, & Prince-Pike, 2008-2013) and IBM CPLEX (Ding, Wang, Dong, Qiu, & Ren, 2007), (Goetschalckx, Vidal, & Dogan, 2002). More recently, the limitations of Frontline Solver Standard package that is shipped together with Microsoft Excel have been overcome with the release of freeware add-ins like OpenSolver (Perry, 2012), (Aeschbacher, 2012) which are capable of solving linear models of almost unlimited size.

The problem that is analyzed and solved, through a series of different methods in this paper, is the redesign of the logistic network of a Brazilian company, from the consumer packaged goods sector (CPG), with headquarters and factory located in São Paulo city and a customer and distribution network spreading over São Paulo state. Its solution was part of a consulting project engaged by the undergraduate students company from Escola de Adminstração de Empresas de São Paulo (namely EAESP/FGV's

Empresa Júnior). EAESP/FGV is the leading business school in Brazil. From now on the CPG company which is the study object of this article will be just called "company". The objective is to minimize the overall distribution costs through the adequate choice of distribution centers, DCs ("centros de distribuição" as they are called in Portuguese), transport routes from factories to DCs and the appropriate assignment of customers (mainly wholesale companies and supermarkets) to each DC, based on demand and costs levels.

At first the company decided to choose the DCs only by the criterion of proximity from its customers. Due to volume increase, coupled with stiff competition, the logistics costs started representing a considerable percentage of the company profits. This prompted the upper management to try alternatives to the selection process, which would be based not only in one but in several factors. It was hoped that this way, besides getting an optimal solution for the problem at hand, the model could let managers think about the implications of their decisions allowing a continuous improvement process to be deployed, and letting the spread of this quantitative based decision process to be spread over other regions.

Another factor that runs in parallel with the analysis is the environmental implications of this redesign. Since the model aims a total cost reduction and the problem at hand deals with the movement of goods by carretas, trucks, and other kinds of diesel vehicles, there's a carbon dioxide emission reduction that could also have a net positive impact in the company's results.

When you face the redesigning problem of a logistics network there are five elements and their relationships that must be considered: suppliers, factories, DCs, wholesalers and customers (Chopra, 2003), as can be seen in Figure 1.

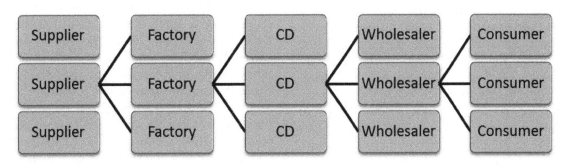

Figure 1 - Logistics Network Elements

Since there's just one factory in the company's structure and the company itself doesn't sell directly to consumers, the model from Figure 1 was simplified from four levels and five different elements to two levels and three different elements (as can be seen in Figure 2) when applying it to this specific problem.

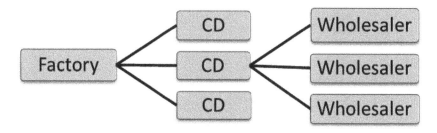

Figure 2 - Company's Model Logistic Network Elements

In order to minimize the total operational cost of a network, the relationships among the elements and their constraints are modeled with linear functions. This is done in order to guarantee the existence of only one optimal solution or no solution at all. The decision variables are the DCs location, the customers that will be assigned to each DC and the transportation routes that will be chosen to fulfill the customer's demands. The last cost factor is the fixed cost to operate a DC, which due to its nonlinear relationship with the amount that will be moved (since it can be either zero or a fixed amount) requires a linearization procedure in the modeling (Sitek & Wikarek, 2012).

In the next section it will be presented a brief review of network modeling and optimization procedures as they relate to the process of optimizing and redesigning a logistic network.

2. LITERARY REVIEW

The area of logistics optimization through linear programming methods, have undergone a strong development, especially after the 80's (Sitek & Wikarek, 2012). From an historical perspective the optimization of goods transportation had been studied as early as 1930, as part of the USSR railway network management (Schrijver, 2002). Its development had a major boost in the late 40's, with the development of the Simplex Method by Dantzig and its application to various problems either in specific engineering application or in the solution of broad classes of managerial problems (Dantzig, 1963) and another one with the development of combinatorial optimization (i.e. integer programming methods) in the beginning of the 60's (Schrijver, 2005). Due to the linear

relationship between transportation cost, distance and transported weight, around the problem of minimizing the cost of moving products from factories to distribution centers and there to customers have evolved a whole class of different solutions, each tailored to a specific piece of the logistic network (Berwick & Mohammad, 2003), (Kropf & Sauré, 2011), (Nagourney, 2007).

On a theoretical point of view, a network system consists of a series of nodes interconnected by arcs representing the transportation routes available (Nagourney, 2010). As in a real distribution network there are nodes which supply products to the network and there are nodes which demand them. The challenge is to move the products from the supply nodes to the demand nodes in the least costly possible way (Tsao & Lu, 2012) . Most of these problems can be solved by assigning a different variable cost to each arc in the network, supposing that the amount to be moved in each route are the decision variables and trying to minimize the linear combination of amounts to be moved and variable costs in each arc. This solution must satisfy a series of more or less standard constraints . The amount to be moved out of a supply node must not exceed the amount available to be moved and the amount to be moved into a demand node should at least satisfy it (Hockey & Zhou, 2002). The flux of products in each intermediary node must be kept smooth, in other words, the amount coming to the node must equal the amount that leaves it minus the amount of products that will remain in the node, besides that, the arcs can be submitted to a maximum flux constraint (Cui, Ouyang, & Shen, 2010). Finally to simplify things when the problem is being deployed in a spreadsheet, the supplies are considered to be negative values (as opposed to the demands, which will be considered positive). This allows all flux constraints in the nodes to be thought of having the following structure:

Arrivals – Departures >= Supply(-) or Demand(+)

If the optimization problem under development requires also decisions regarding the availability of a specific network structure (like having available or not a distribution center or a specific route), binary variables can be used to model this kind of decision. As long as these binaries variables are kept adding or subtracting their values within each other, the problem will be kept linear and so, entitled to have an unique solution which will be able to be found by the simplex method (Altiparmak, Gen, Lin, & Paksoy, 2006).

Besides the Simplex Method, integer programming problems (when the decision variables must be found among the integers) can be addressed through the Branch-and-Bound algorithm (Land & Doig, 1960) (Yuan, 2013), and the selection of different options can be modeled through the usage of binary variables, which are in fact a subset of the integer programming methods (Winston, 2003) (Amin, 2012)

All this theory must now be coded in some programming language and presented to a computer so the specific problem at hand can be solved. Through the years, since the development of the simplex method, the computer has been used in this kind of problem, first by programming each problem in a high-level broad purpose programming language like FORTRAN or C (Press, Teukolsky, Vetterling, & Flannery, 2007) , than in specific software packages, tailored to solve linear programming problems like LINDO (Copado-Mendez & Blum, 2013) and finally as a toolbox embedded in another software designed to make calculations in general like Microsoft Excel (Ravindran & Warsing Jr., 2013) or Mathworks MATLAB (Ko, Tiwari, & Mehnen, 2010) . It was as an embedded package in Microsoft Excel, called Solver and developed by Frontline Systems, that LP problems and the Simplex Method achieved widespread usage, especially in non-technical audiences (Godfrey & Manikas, 2012). The problem that will be described in this paper is presented and solved in a simplified version through problem modeling in a spreadsheet using Microsoft Excel and Frontline Solver. The full scale problem is solved by three different procedures. The first procedure is a com-

bination of Matlab, Opti Toolbox (Currie & Wilson, 2012) and CPLEX (Goetschalckx, Vidal, & Hernandez, 2012). The second one is done by using OpenSolver Add-in and the third one by calling CPLEX from within Microsoft Excel as if it was an Add-in.

3. METHODOLOGY

As stated in the previous section, the problem that will be analyzed here was solved as part of a consulting project developed by EAESP/FGV's Empresa Júnior. Therefore the project was conducted from the onset as a case study of logistic network optimization through the usage of linear programming.

The consultants (business administration undergraduate students from EAESP/FGV) conducted a series of meetings with the company's upper management to understand the strategic choices available to them when redesigning and optimizing company's logistic network and with middle-level managers to understand which kind of problem they were facing, due to the lack of optimization on their network.

In fact, there happened a total of eight meetings. Three of them with the upper management, one of them (the first one) an introductory meeting where EAESP/FGV's Empresa Junior was presented, followed by a second meeting where the upper management described what they would like to get from the project in terms of managerial decision support and the third one to present the results, upon project completion. In these three meetings the company's national logistics and distribution director and two of his senior staff members were always present.

The other five meetings were conducted among the mid-level managers at the company's distribution center in São Paulo, were they presented to Empresa Junior's consultants the companie's distribution network, their decision system (which was in fact a best guess system) and provided the consultants with the information needed to develop and deploy the optimization model,

After this series of meetings (which includes the last meeting with upper management), the consultants presented to the company's managers (in three workshops) a series of choices, not only to optimize the structure in place but also to give advice on how to modify the network structure in order to better manage it. All these choices were backed up by quantitative analysis done with linear programming and made possible by the flexibility and power of the packages above mentioned. Although the free version of Solver that comes bundled with Microsoft Excel is not capable of handling more than one hundred and fifty variables and constraints, Microsoft Excel can manipulate matrices of a huge size. Besides that Mathworks Matlab, IBM CPLEX or OpenSolver can handle an almost unlimited number of variables when solving a problem through the Simplex Method.

The specific problem that will be first described and solved is a simplified version of the original one. This way it can be reproduced and the results presented here replicated using only the Standard Solver. The full scale problem will require a different software packages and when being solved by Matlab an entirely different approach than just creating a spreadsheet depicting the overall calculations that are being made, and these processes will also be described in detail. After that, the models will be solved in a number of different scenarios, so the solutions, methodologies and managerial options can be presented and properly evaluated.

The objective of the model is to minimize the total logistic cost when moving goods from the company's main distribution center to cities in São Paulo, Minas Gerais and Paraná states, where the retailers are located. This model has three levels : a factory (which is considered the main distribution center), distribution hubs (DCs) and retailers. This means that the products are first moved from the factory in São Paulo to the distribution hubs. There the products undergo a cross docking process and then are moved to the retailers. This model therefore is focused in the supply chain middle levels, ignoring suppliers and consumers. The objective function is composed of three costs: a) the cost of moving the goods from the factory to the transportation hubs, b) the operational cost of executing the cross docking in the transportation hubs themselves and c) the cost of moving the goods from the transportation hubs to the retailers. The decision variables are the number and location of the distribution hubs and the retailers that will be assigned to each distribution hub. The model constraints are the minimum and maximum quantities of products that can be transported from a distribution hub, and the demand of goods in each retailer. As parameters, the model uses: a) the distance between each pair of city, b) the vehicle capacity and c) the cost per km per ton moved, which depends on the transportation modal being employed and the retailer demand.

From a purely algebraic point of view the model can be represented as follows:

Variables

X_{ij} = amount of goods moved from transportation hub j to retailers located in the city i.

T_j = transportation hub in location j. Binary variable, if city j is chosen to host a transportation hub, $T_j = 1$, if not $T_j = 0$.

D_i = quarterly demand of retailers in city j.

H_j = unit cost (in Brazilian Reais per km per ton) of goods moved from distribution center (factory) to transportation hub located in city j.

Since the unit cost depends on the modal used to move goods from one point to another and the movement of goods from the factory to the transportation hubs is done only by large trucks (*carretas* in Brazilian Portuguese), this cost is fixed in R\$ 0,714 per km per ton moved. This amount is based on historical data provided by the company and is the average of several different transportation costs with fully loaded large trucks (28 tons) on different distances, ranging from fifty to two hundred and fifty kilometers.

C_{ij} = unit cost (in Brazilian Reais per km per ton) of goods moved from transportation hub located in city j to retailers in city i.

In this case, the unit cost will depend on the transportation modal used to move goods from transportation hub j to retailers in city i. These costs can be: a) R\$ 1,865 for vans, which will be used for demands of less than two tons of goods per month, b) R\$ 1,557 for "3/4 s", which will be used for demands of more than 2 tons and less than 3,5 tons a month, c) R\$ 0,810 for "*tocos*", which will be used for demands of more than 3,5 tons a month and less than 7 tons and d) R\$ 0,851 for "*trucks*" (small trucks, not to be confused with large trucks which are called *carretas*) which will be used for demands of more than 7 tons a month. These *trucks* have a capacity of 12 tons.

It could be argued that demands of more than 12 tons a month (which are the capacity of a *truck*) should be fulfilled using a *carreta* (which has a capacity of 28 tons), but this is not the case. Maneuvering a *carreta* within a city neighborhood is a tough task (sometimes it is not possible), so *trucks* must be used, even if it is needed more than one *truck* to fulfill the task at hand.

M_j = movement cost per ton within the transportation hub itself. This is the cost of cross docking and will be considered R$ 60,00 per ton moved.

T_{min} and T_{max} = for a city j to be considered as a potential candidate for hosting a transportation hub a minimum amount T_{min} must be moved (i.e. cross docked) within the transportation hub. On the other hand T_{max} will be considered in this analysis a huge number in order to do not pose a real limit on the amount that a transportation hub will handle.

Objective function

Supposing that there're m cities to fulfill some demand n cities that can possibly host a transportation hub, the expression to be minimized will be:

Total Cost = Transp1 + Mov + Transp2

Where:

$Transp_1$ = total cost of moving goods from the distribution center (factory) to transportation hubs

$$Transp_1 = \sum_{j=1}^{n} H_j X_{ij}$$

Mov = total cost of cross docking in the transportation hubs

$$Mov = \sum_{j=1}^{n} M_j \sum_{i=1}^{m} X_{ij}$$

$Transp_2$ = total cost of moving goods from transportation hubs to retailers

$$Transp_2 = \sum_{i=1}^{m} \sum_{j=1}^{n} C_{ij} X_{ij}$$

This objective function has to fulfill the following constraints:

The amount of goods moved from a transportation hub to a specific city must satisfy its demand

$$\sum_{j=1}^{n} C_{ij} X_{ij} \geq D_i$$

The amount of goods moved out of a transportation hub must be equal or higher than the minimum amount established for a city j to host a transportation hub, otherwise the amount has to be equal to zero.

To keep the problem linear, T_i is a binary variable and T_{min} is a fixed amount. If T_i is 1 the constraint guarantees a minimum amount of goods to be moved (i.e. cross docked), if T_i is 0, the amount will be 0 because this is a minimization problem.

$$\sum_{i=1}^{m} C_{ij} X_{ij} \geq T_i * T_{min}$$

The simplified version of this model, which will be optimized by Frontline Solver, can be seen in figure 3.

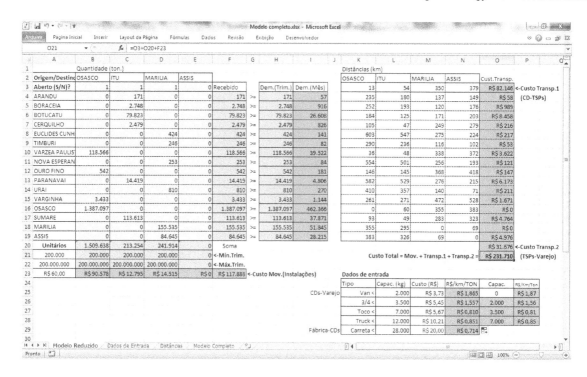

Figure 3 - Logistics Model - Simplified Version

In this version, there are four possible locations to be chosen as transportation hubs and sixteen locations where the demand must be fulfilled. The values presented above are already the optimal solution values for the simplified model. This version was developed so the modeling process errors could be easily located. Once the reduced model version started giving correct results, the model structure was deployed over the entire volume of possible DCs and retailers locations.

The complete model on the other hand deals with four hundred and sixteen cities where the demand must be fulfilled and thirty three locations to choose as possible candidate for a transportation hub. Therefore it has to handle approximately fourteen thousand variables. Due to its size, for purposes of presentation, it is divided in three spreadsheets, one for parameters, another just for the distances matrix and another for the model itself, as can be seen below. It must be noted that to show the whole tables (i.e. spreadsheets), several lines have been hided from view.

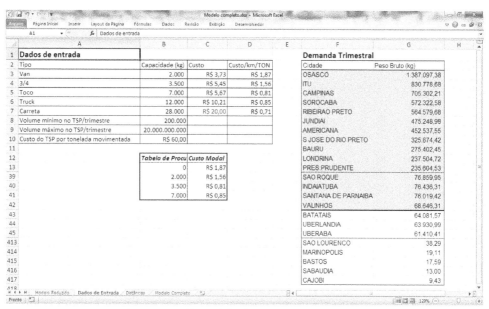

Figure 4 - Full Model - Parameters Spreadsheet

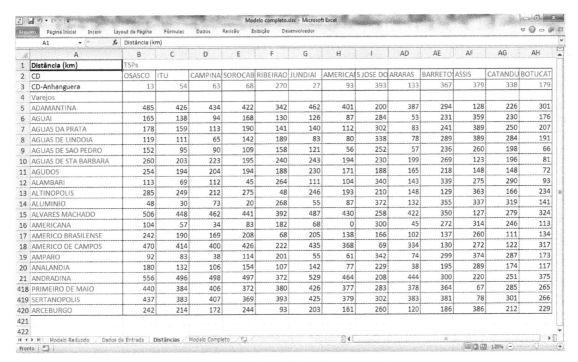

Figure 5 - Full Model - Distance Matrix

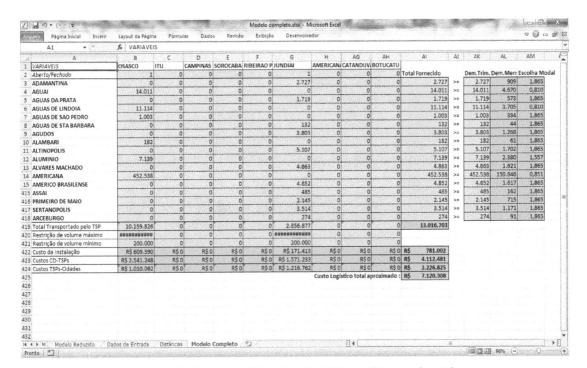

Figure 6 - Full Model - Objective, Variables and Constraints View

As stated earlier, to deal with the full scale model, an industrial class optimization environment is required. Three choices will be used to optimize the full scale model: a) Matlab, Opti Toolbox and Cplex, b) Excel, Cplex and Cplex Excel Connector and c) Excel and OpenSolver Add-in.

The first option is Matlab, Opti Toolbox and Cplex. The main reason is the numerical features and availability of Matlab (especially in academic environments) and the possibility of using Opti Toolbox and CPLEX in their academic versions without any performance constraint.

Matlab is an environment that was originally developed to handle matrices, and to use it effectively everything must be converted to this modeling paradigm. As an example suppose that one wants to solve the following optimization problem:

Find $\mathbf{x}>=0$ that minimizes : $f(x) = -5x_1 - 4x_2 - 6x_3$,

subject to :

$$x_1 + x_2 + x_3 <= 20$$
$$3x_1 + 2x_2 + 4x_3 <= 42$$
$$3x_1 + 2x_2 <= 30$$

In Matlab it would be necessary to issue to following set of commands :

$f = [-5; -4; -6]; A = [1 -1 1; 3 2 4 ; 3 2 0]; b = [20; 42; 30];$

$lb = zeros(3,1); [x,fval,exitflag,output,lambda] = linprog(f,A,b,[],[],lb); x$

This happens because the problem to be solve (i.e. optimised) needs to be put in matrix form A.$\mathbf{x}$ <= $\mathbf{b}$ (the constraints) and f.$\mathbf{x}$ (the objective function). After that the function *linprog* is called to produce the output vector $\mathbf{x}$ with the solution. The function *linprog* (as most Matlab optimization functions) requires the matrices f, A, b (as described above), A_{eq}, b_{eq}, if there are constraints of the form Aeq.$\mathbf{x}$ = $\mathbf{b}_{eq}$ (equality constraints), and lower and upper bounds vectors, lb and ub (if there are such limits) for the solution vector $\mathbf{x}$.

The outputs are: the solution vector $\mathbf{x}$, the value of the objective function f, named fval, the output flag (+1 if the process finished ok, -1 if not, etc…) and some process and optimization parameters that are of no interest for this paper.

This example illustrates the standard that the LP problem structure must adhere to, so it can be solved in Matlab. If, for example, the problem requires the variables all to be of binary type, another solver should be called, *bintprog*, but the overall structure of the parameters and outputs remains the same.

Next, it is needed to include a mix of continuous and binary variables in the problem, something that the standard optimization packages of Matlab are not able to do. To overcome this limitation it was used the Opti Toolbox, a free Matlab Toolbox developed by the Industrial Information and Control Centre from the Auckland University of Technology, New Zealand. This toolbox can be freely downloaded from   http://www.i2c2.aut.ac.nz/Wiki/OPTI/index. php/DL/DownloadOPTI and includes a variety of solvers, but most important than that, as will be seen later in this section, is the ability to encapsulate several other solvers in its command framework. This way, the standard optimization functions of Matlab for instance, can be called as *opti_linprog* and *opti_bintprog*.

The Opti Toolbox has a function called *opti_lpsolve* that is called with the exact same parameters as described for *linprog* above, plus an extra parameter that describes if the variable will be of continuous, integer or binary type. This parameter is just a string with the chars "C", "I" and "B" listed in the same order as the variables appear in the objective function. Therefore if the variable is not normally present in the objective function expression, a 0 (zero) must be added so it can be "understood" by the function that it "exists".

In the picture below it can be seen a MILP (Mixed Integer Linear Programming problem) structure in a spreadsheet, where the matrices have been prepared to be passed to *opti_lpsolve*.

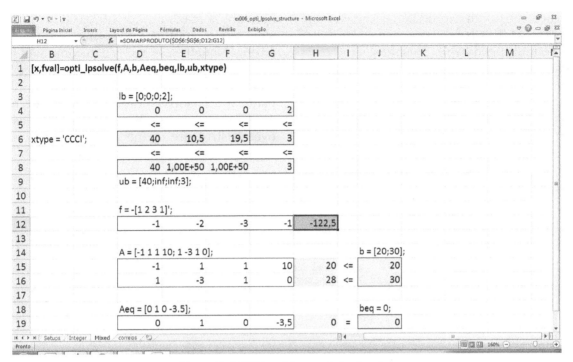

Figure 7 - MILP problem matrix structure

As can be seen above, even if Microsoft Excel is not going to be the engine to solve the problem, nonetheless it is an excellent environment to structure the matrices and have them passed to Matlab and Opti for optimization. To automate this process it were used two functions of Matlab that allow the reading and writing of spreadsheet chunks, *xlsread* and *xlswrite*. Besides that, to avoid the repetition of commands each time the model is being tested, it was created a .m file (example01.m), that reads and writes to the spreadsheet depicted above.

The next step is to try solving the simplified problem through Matlab and Opti Toolbox. To translate the structure as developed in the spreadsheet (to be optimized through Solver) to the matrix structure required by *opti_lpsolve* the constraints must be modified in the following way:

1. *Sum of Quantities from DCs to Retailer $>=$ Retailer Demand*

   This has to be redesigned to:

   $-$ *Sum of Quantities from DC to Retailer $<=$ $-$Demand*

2. *Sum of Quant. from DC to Retailer $>=$ Minimum Volume * DC(binary)*

   This has to be redesigned to:

   $-$ *Sum of Quant. DC to Retailer $+$Min.Volume * DC(binary) $<=$ 0*

3. *Sum of Quant. from DC to Retailer $<=$ Maximum Volume * DC(binary)*

   This has to be redesigned to:

   $+$ *Sum of Quant. DC to Retailer $-$ Max.Volume * DC(binary) $<=$ 0*

Also, in each constraint expression, every variable must be present (if it is not supposed to be there, it has to be multiplied by 0). This makes the coefficient matrix **A** huge and sparse. Besides that, the *Maximum Volume* parameter, since it multiplies a binary variable and is then compared with variables values that can be much smaller than itself, has the potential to make the solution numerically unstable. In fact, the Opti Toolbox was capable of

handling a maximum of 4.000.000 for this parameter, before starting producing results that were wrong. Nonetheless, using a maximum of four million for the maximum volume that can be handled by a DC still produces a valid result. On the other hand, if this solver is to be used in the full scale model this volume should be much greater. The overall structure of the problem matrices can be seen in the following picture:

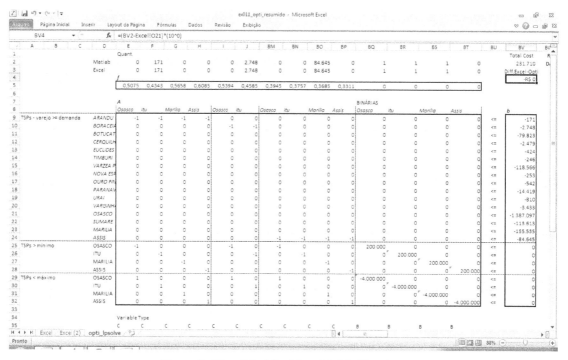

Figure 8 - Simplified Problem Matrix Structure

As can be seen, the spreadsheet structure is correct, since the solutions from Matlab – Opti Toolbox and Excel – Solver are in agreement with each other. Now the problems that remain are: how to overcome the scaling problem and how to replicate the matrix structure presented in the picture above in the full scale model, where the number of columns and lines in the matrix will be incredibly high.

The scaling problem was solved in a twofold way. First, the total demand of goods in each retailer was divided by ten, something that could be done only because de model is linear. Therefore, the results presented by the spreadsheet must be multiplied by ten, so one can know the real quarterly logistics cost. Second, the solver was changed from the Opti Toolbox standard solver to IBM's CPLEX. Opti Toolbox, as noted before, has the ability to encapsulate other solvers, from different vendors, and this is no different with CPLEX. All it is needed to be done is to install IBM's

CPLEX and let Matlab know where it is installed. From that moment own, if someone wants to use the powerful IBM solver, he/she has just to call it using the function wrapper *opti_cplex* with the parameter structure : *[x,fval,exitflag] = opti_cplex([],f,A,[],b,lb,[],xtype).* CPLEX solver was able to optimize the simplified model with a maximum value on the volume that could be handled by a DC as high as two billion tons.

The matrix structure was deployed in a somewhat automatic manner, by using Excel's functionalities of copy and paste to create the sparse matrix A. The resulting spreadsheet can be seen next (take special attention to the number of columns that matrix A has, it spreads over almost fourteen thousand columns, one for each variable, to be more precise thirteen thousand, seven hundred and sixty one columns and four hundred and eighty one lines, making it a matrix with approximately six point seven million elements, most of them zeros, since it is a sparse matrix):

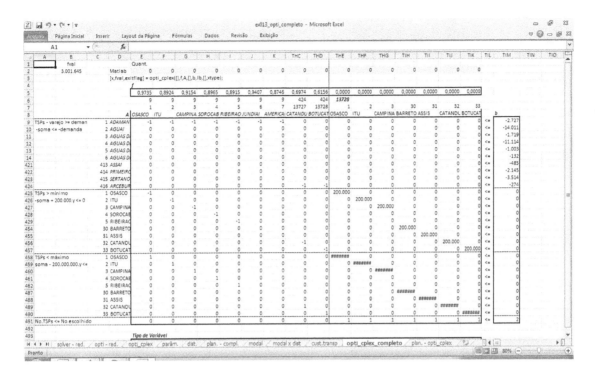

Figure 9 - Full scale problem matrix structure

The spreadsheet with the full scale problem matrices became a file with almost 50Mb. Since the model would have to be optimized for a series of different scenarios, to streamline the process the matrices were saved as separated .csv files before being read by Matlab.

Since the company wanted to analyze the effects of having from one to thirty three DCs, the problem had to be solved for all these different scenarios, and the solutions recorded one by one. To accomplish this task it was created another .m file (report.m), which called repeatedly the optimization procedure, generated the solutions and recorded them in another spreadsheet, that were later analyzed by the company's management team.

The major drawback of this solution method is its complexity. This is due to the fact that Matlab, Opti and CPLEX require the matrices to be passed to them in a standard format. Of course, Matlab is an environment that can deal both with linear and non-linear problems of arbitrary complexity and this is a major advantage. On the other hand, one must consider that this problem in particular and logistics in their great majority are of linear kind. Besides, the solution core was done by CPLEX. Matlab and Opti were used as a means to access CPLEX functionality.

It is natural at this point to ask if it would not be possible to use CPLEX as a standalone program. This

is true, but CPLEX has its own complexities. What would be nice is to have the flexibility of the spreadsheet environment coupled with the power from CPLEX engine. This can be achieved with CPLEX Excel Connector and is the second option that is analyzed here.

The academic version of CPLEX can be downloaded from: http://www-03.ibm.com/ibm/university/academic/pub/page/ban_ilog_programming. The product installs both Matlab and Excel connectors. To make CPLEX Excel Connector available in Microsoft Excel the user should first go to *Files, Options, Add-In, Go* and the choose *Search* from the menu. Than it should locate the file *cplex125.xll* which usually stays in the directory: C:\Program Files\IBM\ILOG\CPLEX_Studio125\cplex\excel. After that, there will be a menu in the ribbon Add-In, called CPLEX. Clicking on it, it will be opened a window for parameters, variables, objective and constraints input.

It also has features similar to Frontline Solver, although some care must be taken when dealing with constraints. To show how a model can be optimized through CPLEX Excel Connector let's optimize the MILP example depicted in figure 7, which was structured to be transferred to Matlab. The model and CPLEX Excel Connector interface can be seen in the next figure, below:

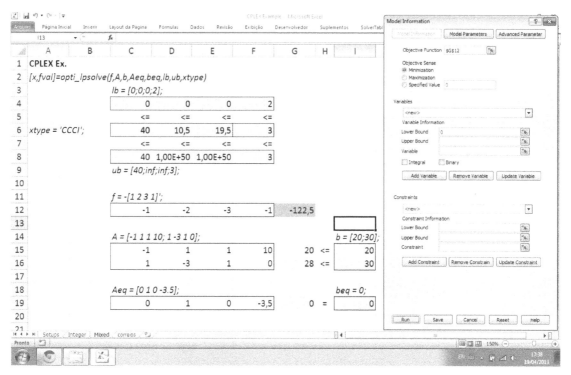

Figure 10 - CPLEX Excel Connector Interface

As can be seen in the picture above, instead of transferring the matrices to the solution engine, one only needs to express the relationships in the spreadsheet and have the cell's addresses transferred to the interface. As stated above, some care must be taken when modeling the constraints and variables. The variables can have lower and upper bounds, but they must be passed together with the variables cells themselves and not as a separated constraint. The interface has a minor 'bug', which is, the integer variable mark is written as 'integral' but this doesn't seem to affect the performance of the environment. About the constraints, CPLEX must receive fixed numbers as lower and upper bounds. The variable cells are the constraint cells themselves. Therefore if one wants to deploy in the spreadsheet a constraint like $x_1 < x_2$,

where x_1 and x_2 are variables themselves, this constraint has to be deployed in a cell as $x_1 - x_2 < 0$. If there are no lower or upper bound, there's no need to indicate that in the interface. Leaving the field as blank will make CPLEX assume that the constraint (or the variable) is unbounded. Besides that, all the values that are going to be needed to solve a model (be they variables or parameters), must be placed in the same sheet. As will be shown later in this section, CPLEX Excel Connector was not able to handle a model when a parameter as being transferred from another sheet. Therefore, when solving the full scale model, the distance, demand and parameters matrices will have to be put next to the model itself.

The next figures depict the spreadsheet full scale model deployment.

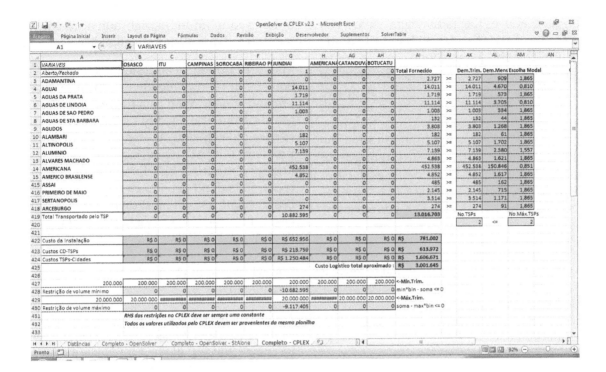

Figure 11 - CPLEX Excel Connector Full Scale Model Deployment

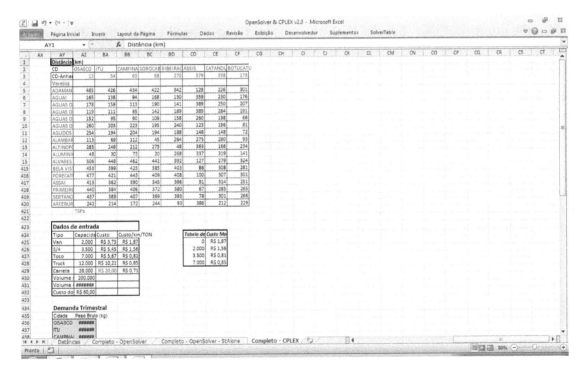

Figure 12 - CPLEX Excel Connector Full Scale Model Parameters

As can be seen above, the parameters have been transferred to the same spreadsheet that will handle the model itself. If this is not done, CPLEX loops around the problem without being able to stabilize it and never finishing its job. When this happens, pressing ESC or CTRL+C keys will be of no avail, having the user to shut Excel itself down by pressing CTRL+ALT+DELL and selecting Excel from the Windows Task Manager. Although this is something that should be addressed by IBM, as long as the user saves its spreadsheet before running CPLEX, there's no problem with the file and the procedure can be started again.

The performance is impressive. Starting from a null solution (where all variables are set equal to zero) it sets up the model and optimizes the model in less than one minute and thirty seconds, when using a modest Sony Vaio laptop with Intel Centrino with 2 cores of 2.1GHz and 3GB of RAM memory.

On the downside, although stated by CPLEX's documentation that CPLEX engine can be called by VBA, the methods are complex and are not well documented, so it was not possible to generate automatically the report with the minimum cost based on the number of DCs. Besides that, although impressive in its speed and sophistication when handling linear problems, CPLEX is not well suited for non-linear problems. It can only handle quadratic problems and the software is designed to test the problem looking for other kinds of functions, which when found made CPLEX stop the optimization process.

So far, CPLEX engine coupled with its Excel Connector has the speed and flexibility needed to be used by a manager on a daily basis but it lacks easy VBA programming features and of course, if it is to be used in an industrial (non-academic environment) it will have a non-negligible cost.

The third option to be tested is OpenSolver. OpenSolver can be downloaded from http://opensolver.org and is developed as a freeware by Prof.Andrew Mazon from The Faculty of Engineering from University of Auckland. It also can handle only linear problems but has an interface more similar to Frontline Solver than CPLEX Excel Connector, where the constraints don't need to be put in a specific format. Besides that, as will be shown in this section, it can easily be called from VBA. More important than that, it can handle an almost unlimited number of variables and constraints, making it an excellent option when one have to deal with linear problems of considerable size.

To install it, after downloading the .zip package from the authors website, all is needed to be done is to unpack the .zip file in a suitable directory and follow the same procedure already followed to include CPLEX Excel Connector Add-In. Usually the default directory for third-party Add-Ins installed by the user itself is the directory C:\Users\gustavo\AppData\Roaming\Microsoft\Suplementos. In this case it was created an specific subdirectory (OpenSolver21) to receive the unpacked files. After that all is needed is to include the file *opensolver.xla* in the Add-In menu of Excel and OpenSolver will be available as part of the Data ribbon.

The model structure itself was pretty similar to the reduced model (only bigger), and the OpenSolver interface also pretty similar to Frontline Solver. As a matter of fact, OpenSolver captures any model that has been developed previously in the spreadsheet by Frontline Solver and can exchange changes with Frontline Solver.

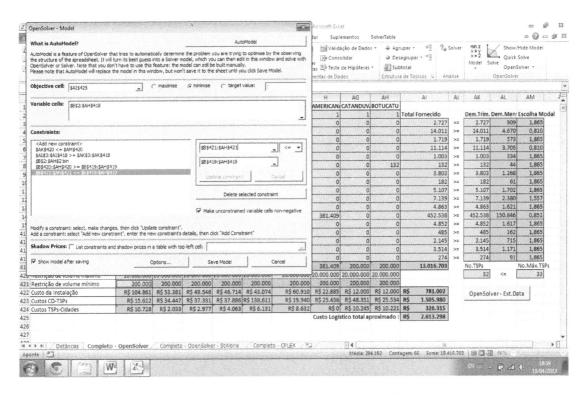

Figure 13 - OpenSolver Parameters Interface

The first time OpenSolver was put to optimize the full scale model (see figure below), the spreadsheet was designed with parameters being passed from other sheets, since this way the problem can be broke in smaller pieces. This way the optimization process took of a model starting from a null solution and having to choose up to thirty three DCs took almost thirty minutes. At first it was taught that this was happening because the algorithms deployed in OpenSolver were less efficient than the CPLEX similar ones. It was noted however that the spreadsheet took a long time setting up de model and, more important than that, used to present intermediate results similar to the ones CPLEX presented when looping without finding a solution. This was an indication that the problem was not in the OpenSolver algorithms but in the model structure itself.

The spreadsheet was then redesigned without the external links. In fact, OpenSolver was applied in the CPLEX sheet which was already running in stand-alone mode (i.e. without external links). Considering that OpenSolver is a freeware developed by an engineering department in an university, seeing it solve the full scale model, with thirty three DCs to choose in the same amount of time that it took CPLEX to accomplish the same task (ninety seconds) was beyond impressive.

The next task was to develop a way to call this model thirty three times, so the full managerial report could be generated. If OpenSolver had equaled CPLEX on the matter of performance here it surpassed its professional counterpart. It was decided to write a small report in three parts of eleven lines. In each part there was going to be put the number of DCs and the total cost obtained after the optimization process had been run. To automate this task it was created a command button that called a small VBA program (the listing can be seen in Appendix A). Of special interest is the easiness with which one can call OpenSolver from VBA. Once the model is designed and objective, variables and restrictions are set in the OpenSolver Model interface, there are only two VBA functions that are required: *initializequicksolve* and *runquicksolve,* which are called from VBA without any special argument. The spreadsheet with the report result can be seen in the next figure.

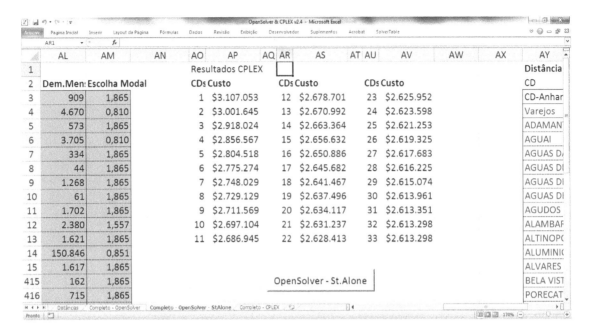

The table in the figure shows:

	AL	AM	AN	AO	AP	AQ	AR	AS	AT	AU	AV	AW	AX	AY
1					Resultados CPLEX									Distância
2	Dem.Men:	Escolha Modal		CDs	Custo		CDs	Custo		CDs	Custo			CD
3	909	1,865		1	$3.107.053		12	$2.678.701		23	$2.625.952			CD-Anhar
4	4.670	0,810		2	$3.001.645		13	$2.670.992		24	$2.623.598			Varejos
5	573	1,865		3	$2.918.024		14	$2.663.364		25	$2.621.253			ADAMAN'
6	3.705	0,810		4	$2.856.567		15	$2.656.632		26	$2.619.325			AGUAI
7	334	1,865		5	$2.804.518		16	$2.650.886		27	$2.617.683			AGUAS D,
8	44	1,865		6	$2.775.274		17	$2.645.682		28	$2.616.225			AGUAS DI
9	1.268	1,865		7	$2.748.029		18	$2.641.467		29	$2.615.074			AGUAS DI
10	61	1,865		8	$2.729.129		19	$2.637.496		30	$2.613.961			AGUAS DI
11	1.702	1,865		9	$2.711.569		20	$2.634.117		31	$2.613.351			AGUDOS
12	2.380	1,557		10	$2.697.104		21	$2.631.237		32	$2.613.298			ALAMBAF
13	1.621	1,865		11	$2.686.945		22	$2.628.413		33	$2.613.298			ALTINOP(
14	150.846	0,851												ALUMINI(
15	1.617	1,865												ALVARES
415	162	1,865						OpenSolver - St.Alone						BELA VIST
416	715	1,865												PORECAT

Figure 14 - OpenSolver Managerial Report

4. RESULTS

Both the simplified and the full versions of the model run smoothly, although the full version required either the transfer of the model to the Matlab environment where the Simplex method could be applied to the huge matrices involved or the usage of Microsoft Excel, CPLEX and CPLEX Excel Connector or Microsoft Excel with OpenSolver Add-In.

The simplified version, using the Simplex Method, required less than 5 seconds when running the Standard Solver. On the other hand, the full scale model required almost 2 minutes for each optimization scenario.

Matlab, Opti Toolbox and CPLEX were the most sophisticated group of software used to solve the full scale model. There's no doubt that this trio has the biggest amount of resources, its performance is excellent and should be considered especially if the optimization procedure would require a non-linear feature. On the other hand, since Matlab has to deal with matrices in an almost raw form and its optimization routines require a fixed input format, it can be cumbersome to generate them, as became obvious in this problem. Therefore, if one is looking for a flexible interface that can be easily manipulated by non-technical personnel, other options should be considered. That is even more the case if the model will be of linear type.

CPLEX and CPLEX Excel Connector are the next choice. Since CPLEX was the real power horse be-

hind the trio Matlab, Opti and CPLEX itself, and IBM has released an Excel Connector it would be an excellent candidate for being the preferred option. This is even more the case because IBM is a name that provides a solid guarantee for whoever decides to use its products. The problem with CPLEX was the difficult way with which one had to deal if trying to call the optimization procedure from VBA. Besides that, CPLEX requires some specificities in the spreadsheet design, especially when dealing with constraints, which make him a not so flexible option.

The third option, and its performance and flexibility, was a big surprise. OpenSolver demonstrated having the same performance as CPLEX and is structured in a way that is as flexible as Frontline Solver itself. Besides that, its VBA interface is simple enough for a non-technical person to use it, as long as he/she has some basic programming knowledge. When you consider that OpenSolver is a freeware, it becomes an obvious choice. Therefore the model and the managerial report were presented to the company's management in Microsoft Excel with OpenSolver Add-In.

The management team wanted to analyze what was the effect on the total logistics cost of increasing from one to thirty three the number of DCs spread in the distribution network. The results are presented below, where the total cost is the total quarterly logistic cost in Brazilian Reais. Full details can be seen in the spreadsheets *"report.xls"* (for the environment

Matlab, Opti Toolbox and CPLEX) and *"opensolver & cplex.xls"* (for the other two environments).

N.DCs	Total Cost	N.DCs	Total Cost	N.DCs	Total Cost
1	R$ 31.070.534	12	R$ 26.787.013	23	R$ 26.260.671
2	R$ 30.016.453	13	R$ 26.710.367	24	R$ 26.235.984
3	R$ 29.180.240	14	R$ 26.633.644	25	R$ 26.212.528
4	R$ 28.565.673	15	R$ 26.566.322	26	R$ 26.194.947
5	R$ 28.045.175	16	R$ 26.508.863	27	R$ 26.176.833
6	R$ 27.752.740	17	R$ 26.456.825	28	R$ 26.162.250
7	R$ 27.480.289	18	R$ 26.414.672	29	R$ 26.150.744
8	R$ 27.291.289	19	R$ 26.374.946	30	R$ 26.139.605
9	R$ 27.115.694	20	R$ 26.341.173	31	R$ 26.133.512
10	R$ 26.971.041	21	R$ 26.312.369	32	R$ 26.133.716
11	R$ 26.871.384	22	R$ 26.284.127	33	R$ 26.133.716

Considering that the company's total logistic cost was approximately R$ 45M, with a total of 5 DCs, based on the results presented above the company would be able to achieve a total cost reduction of about 28%, representing a total of R$ 17M in economies, on a quarterly base, just by redesigning their distribution network.

Other interesting result can be seen if one looks at the total cost reduction from one number of DCs to the next number. If one goes from four DCs to five DCs the total cost reduction is approximately R$ 520.000 per quarter. When increasing the number of DCs from five to six the cost reduction is approximately R$ 290.000 per quarter. If the number of DCs is increased beyond seven, the total optimal cost keeps reducing but the marginal reduction is increasingly smaller. The total number of five DCs is the transition point from more than half a million reais per quarter to almost half of this amount. Somehow, when the management team had first decided to choose in having five DCs they had chosen the number of DCs that would provide the best marginal cost reduction.

5. CONCLUSIONS

Although being a practical application of a theoretical framework, the work presented in this paper can be tough of having one major theoretical implication, which is the possibility of applying linear optimization techniques in large scale within Brazilian companies to achieve large economies of scale in distribution networks. This application can have not only a quantitative impact in the operations of Brazilian companies but also a qualitative impact that could generate new levels of operation efficiency leading to new supply chain models when applying them to emerging markets operations.

Before this project, the company used to choose the DCs locations by following the competition decisions, even knowing that its direct competitor had a product portfolio somewhat different from its own. With the decision process now based also on quantitative data it is now possible to analyze what will be the consequences in terms of cost of choosing one location versus another. The logistics director stated that this has allowed the logistics division to be better integrated in the company's strategic planning and to get a fair share of recognition in the results. The next step is to spread this decision process to other regions in Brazil, especially in other the southern states (Santa Catarina and Rio Grande do Sul).

When it comes to the results that the full scale model produced, they were both reassuring and challenging for the company's management. As can be seen in the Results section, the marginal cost reduction that was calculated with the full scale pointed to a maximum marginal optimal cost reduction in the region of five DCs. Since this was the number that the management team had chosen previously, this fact was highly reassuring for the company's past decisions and for the model validity is general. After that, there was no doubt that the options offered by it were not something "purely theoretical stuff from a group of academics" but real stuff that could be used to really reduce the company's total logistics cost. This fact is reflected in the company's decision to have workshops with middle level managers to explain how this model could be deployed in other regions and in a smaller base or in different contexts. The consultant perception was that suddenly the company's middle level management team had realized that "there's a practical side on this theoretical stuff".

6. LIMITATIONS AND SUGGESTIONS FOR FUTURE RESEARCH

When talking about the scope of this study, it is obvious that it is a very limited application of linear optimization techniques. More work should be done to determine if this case represents a distribution network tipical situation of a company in the Brazilian CPG's sector or if it was a very specific individual case, although one can argue that cases where the improvements are achieved in a way so quick and straightforward as this one cannot be a statistical anomaly. This is even more so if one considers that this company is a leading CPG manufacturer and distributor.

Nonetheless, this study opens some areas for future research. As can be seen from the example presented here, the modeling of a logistics network of a considerable size can be done with spreadsheet resources and its optimization be accomplished with the support of software running in microcomputers.

What is even more interesting, and could be further analyzed in the future, is the amount of optimization that companies in Brazil use when designing their logistic infrastructure. It was a surprise for the consultants of EAESP/FGV Empresa Júnior the interest and willingness to cooperate that the logistics management team have received the results and the efforts that they took to understand how the optimization process was achieved and how it could be improved.

This is an indication that even classroom examples, when properly applied, can have a huge impact in the competitiveness of well established companies of considerable size in Brazil.

Regarding the environmental implications of this network redesign, specifically in this project the total carbon dioxide emissions where not calculated, although it is a matter of logic to think that if one is

reducing the total goods movement cost, there will be a positive impact in the emissions by the company. This is something that management will be addressing in a future analysis, because it can add an impact in the company's results in a way that is not usually offered by a logistics department (where the net impact is felt mainly by cost reductions and not by revenue increase).

REFERENCES

Aeschbacher, B. (2012, 08 23). *Solving a Large Scale Integer Program with Open-Source-Software Master thesis Submitted at the University of Zurich.* Retrieved 04 19, 2013, from University of Zurich: http://ufpr.dl.sourceforge.net/project/opensolver/OtherResources/Aeschbacher%20Masters%20Thesis%20Solving%20a%20Large%20Scale%20Integer%20Program.pdf

Berwick, M., & Mohammad, F. (2003). *Truck Costing Model for Transportation Managers.* North Dakota, USA: Upper Great Plains Transportation Institute.

Chopra, S. (2003). Designing the Distribution Network in a Supply Chain. *Transportation Research*, pp. 39(2):123-140.

Dantzig, G. B. (1963). *Linear Programming and Extensions.* Santa Monica, California: The RAND Corporation.

Ding, H., Wang, W., Dong, J., Qiu, M., & Ren, C. (2007). IBM Supply-Chain Network Optimization Workbench: An Integrated Optimization and Simulation Tool for Supply Chain Design. *Proceeding of the 2007 Winter Simulation Conference*, (pp. 1940-1946).

Eshlaghy, A. T., & Razavi, M. (2011). Modeling and Simulating Supply Chain Management. *Applied Mathematical Sciences*, pp. vol.5, no.17, 817-828.

Geunes, J., & Panos, P. M. (2005). *Supply Chain Optimization.* New York, NY: Springer Science.

Goetschalckx, M., Vidal, C., & Dogan, K. (2002, Nov 16). Modeling and design of global logistics systems: A review of integrated strategic and tactical models and design algorithms. *European Journal of Operational Research*, pp. Vol.143, Issue 1, pags 1-18.

Huang, C.-H., & Kao, H.-Y. (2012, 12). A Design of Supply Chain Management System with Flexible Planning Capability. *International Journal of Human and Social Sciences*, pp. 5, pags.775-779.

Huang, L. (2012, 07). Modeling and Planning on Urban Logistics Park Location Selection Based on the Artificial Neural Networks. *Journal of Computers*, pp. Vol.7, no.3, 792-797.

Kropf, A., & Sauré, P. (2011). *Fixed Cost per Shipment.* Retrieved March 25, 2013, from http://www.eui.eu/Personal/JMF/PhilipSaure/KropfSaure_2011_12_16.pdf

Land, A., & Doig, A. (1960). An automatic method of solving discrete programming problems. *Econometrica*, pp. 28 (3). pp. 497–520.

McGarvey, B., & Hannon, B. (2004). *Dynamic Modeling for Business Management: An Introduction*. New York, NY: Springer-Verlag.

Nagourney, A. (2007). *Mathematical Models of Transportation and Networks*. Retrieved March 25, 2013, from http://supernet.isenberg.umass.edu/articles/EOLSS.pdf

Perry, K. M. (2012, 04 27). *The Call Center Scheduling Problem using Spreadsheet Optimization and VBA A thesis submitted in partial fulfillment of the requirements for the degree of Master of Science at Virginia Commonwealth University*. Retrieved 04 18, 2013, from Virginia Commonwealth University: https://dizzyg.uls.vcu.edu/bitstream/handle/10156/3746/PerryThesis.pdf?sequence=1

Press, W. H., Teukolsky, S. A., Vetterling, W. T., & Flannery, B. P. (2007). *Numerical Recipes 3rd Edition: The Art of Scientific Computing*. Cambridge University Press.

Radhakrishnan, P., Prasad, V. M., & Gopalan, M. R. (2009, Jan). Inventory Optimization in Supply Chain Management using Genetic Algorithms. *International Journal of Computer Science and Network Security*, pp. vol.9, no.1, pags 33-40.

Ragsdale, C. T. (2008). *Spreadsheet Modleing and Decision Analysis, 5th Revised Edition*. Mason, OH: Thomson South-Western.

Sitek, P., & Wikarek, J. (2012). Cost optimization of supply chain with multimodal transport. *IEEE Processdings of the Federated Conference on Computer Science and Information Systems*, (pp. pp.1111-1118).

Wilson, D. I., Young, B. R., Currie, J., & Prince-Pike, A. (2008-2013). *Industrial Information & Control Centre (I2C2) Home Page*. Retrieved 04 17, 2013, from Industrial Information & Control Centre (I2C2): http://www.i2c2.aut.ac.nz/

Winston, W. (2003). *Operations Research: Applications and Algorithms*. Pacific Grove, CA: Duxbury Press.

Winston, W., & Albright, S. (2009). *Practical Management Science, 3rd revised ed.* Mason, OH: South Western - Cengage Learning.

Gustavo Mirapalheta is Electrical Engineer (UFRGS), Dsc in Business Administration (EAESP/FGV) and prof.of Quantitative Modeling and Decision Analysis at EAESP/FGV. Gustavo's research interests lay in the areas of Operations Research, Simulation and Big Data. Currently works as business diretor at Inventive Solutions, a sales optimization and improvement consulting company. Previously worked as software director at Sun Microsystems and business manager at IBM Brasil

Flávia Junqueira de Freitas is majoring in Business Administration at Fundação Getulio Vargas (FGV-EAESP). She has experience in Consulting due to her two-year job at Empresa Junior FGV (EJ-FGV). She is currently at A.T. Kearney

5

Dyad Buyer-Supplier and its Relation to Financial Performance

Luis Cesar Mondini
University Center Leonardo Da Vinci
cesar.mondini@gmail.com

Denise Del Prá Netto Machado
Universidade Regional de Blumenau
profadenisedelpra@gmail.com

Marcia Regina Santiago Scarpin
mrs.scarpin@gmail.com

ABSTRACT: This study aims to raise which practices of buyer-supplier dyad are related to the financial performance of the manufacturing industry in Brazil. Based of 174 Brazilian companies and a total of 312 respondents, the analysis use CFA (Confirmatory Factor Analysis) to validate the measurement of constructs and multiple regressions to analyze the relationship between practices of buyer-supplier dyad and financial performance. Our results showed a positive relationship dimension--strategic sourcing and buyer-supplier relationships--with the company's financial performance. However, the supplier evaluation system showed no statistically significant relationship. The findings reinforce the lack of formalization in the buyer-supplier relationship, a situation that often hinders the development of a long-term relationship. The lack of systematic evaluations of supplier performance can generate insecurity in the relationship, since historical actions taken by suppliers could serve as a criterion of choice in a future negotiation. As practical contributed to this study, it is expected that managers understand the importance of buyer-supplier relationship to the company's financial performance.

Keywords: dyad; buyer-supplier; financial performance relationship; industry

1. INTRODUCTION

The positive influence of purchasing integration on manufacturing performance confirms anecdotal evidence about the importance of this competence. The purchasing's influence on final product changes, based on acquisition costs and availability analysis, technology forecasts, and supply base capability evaluations. This fact indicates the relevance of purchasing participation in strategy formulation and the need to focus on strategic performance metrics in purchasing performance evaluation and compensation systems. Companies that invest in practices to achieve this integration can expect to see commensurate gains in strategic performance (Narasimhan & Das, 2001).

Practices such as strategic purchasing, supplier evaluation systems, and buyer–supplier relationships are in fact important with respect to the firm's financial performance (Carr & Pearson, 1999). Narasimhan and Das (2001) claim that as in additional practices, supply base leveraging, buyer–supplier relationship development, and supplier performance evaluation, it is necessary to elevate purchasing into a strategic function. These practices should be accompanied by internally focused purchasing initiatives that enable purchasing to become a part of the business planning process to guide to a high level of competitive advantage.

Therefore, this study aims to raise which practices of buyer-supplier dyad are related to the financial performance of the manufacturing industry in Brazil. The central argument is that a well-aligned buyer–supplier relationship would have a measurable impact on the financial performance of the Brazilian manufacturing industry. The manufacturing industry was chosen for two reasons: first, for its relevance in the dynamics of the Brazilian economy (FIESP, 2014); second, for the importance of the role of relations between actors in the supply chain for the sector's competitiveness, since it depends on its suppliers to receive raw materials, development of new technologies, and, in some cases, distribute products in a timely fashion with minimal costs. Another factor is the lack of studies that address the dyad in the manufacturing industry as a whole, well as its relationship with financial performance.

Recent research has stressed that purchasing and supply management can have a profound impact on firms' financial performance (Ellram & Liu, 2002). But, according to Terpend et al. (2008, p. 43), there is little research on the financial factors affected by suppliers and supply chain performance indicators, and the "link between collaborative buyer–supplier relationships and operational performance was an important research focus throughout the past two decades, but the focus on financial performance only became evident after 2001". The authors were surprised when they found in six articles of the 151 reviewed there were dyadic buyer–supplier studies. They recommend more dyadic studies and future research efforts to recognize the limitations of a single theoretical perspective and adopt a multidimensional view to explain how buyer practices and the influence of buyer-supplier mutual efforts help the firm's performance.

The purchasing department possessed a strategic role in the organization (Pearson, Ellram & Carter, 1996; Carr & Smeltzer, 1997), especially in the relationship between buyer and supplier. It is through the relationships that are established with suppliers that inevitably will impact the company's financial performance. The perspective of relationship is opposed to models that perceive the supply a simple transactional exchange. This perspective, the social context of companies that negotiate together is based on reciprocity, cooperation and collaboration, is demonstrating to be an efficient means in relation-specific assets, knowledge sharing, complementary resources/capabilities, effective governance, and a possible source of competitive advantage (Dyer & Singh, 1998).

This study expects to stimulate research on the buyer-supplier relationship, a recent topic which needs further study for comparisons and analysis, as it is still under construction and debate, providing an empirical contribution. The article is organized into more four sections: the second presents a literature review; in the third, the methodological procedures employed to conduct the research are described; the next section presents the data analysis, and finally, concluding remarks are exposed with suggestions for future research.

2. REVIEW OF LITERATURE

The strategic management of resources, in order to increase the competitive advantage of organizations, came to occupy a prominent place on the agenda of managers, increasing the status of purchasing (Carr & Pearson, 1999). Studies on relationship management in the supply chain emphasize the importance of strategic relationships between buyers and suppliers.

For this kind of relationship the coordination between business partners is essential. The intensity of competition in which a company operates is not a matter of coincidence or bad luck. This intensity of competition is caused by external forces belonging to the industry where the organization belongs to and it is significant and present the entire time, making these organizations seek to devise their strategies in which partnerships are strengthening and relevant options in the market (Porter, 1986). The supply area plays an important role for organizations to achieve their goals of cost savings and increased profitability (Anderson & Katz, 1998).

As a consequence, according to Joshi (2009), organizations have demanded an improvement in the performance of suppliers to enable them to maintain this advantage. There is no denying that when negotiation is based only on price, one of the first things that suffer is the quality of the purchased product. The supplier will seek a reduction in the standard of quality and offer only basic specifications using cheaper materials. It has been shown that the collaborative buyer-supplier relationships results in an acquisition of unique value (Corsten & Kumar, 2005). Thus, companies seek ways to improve their activities, some opting to vertically integrate their own acquisitions efforts and others seek market alternatives more agile and with better performance (Carr & Pearson, 1999).

In this conception, this buyer-supplier relationship must be fostered to achieve a process of competitive synergy, where both plot a horizon of opportunities. In this process, the supply area of a company becomes crucial, because it represents the immediate contact with the suppliers. To Lima (2008), the function of the supply area tends to be more strategic as the importance of acquisitions increases, requiring more attention than planning to trading activities that promote longer-term relationships, supplier development, and reduced total cost and not only perform the activities of rebuilding material for immediate stock.

2.1 Strategic sourcing

Carr and Pearson (2002) define strategic sourcing as the process of planning, evaluating, implementing, and controlling highly important sourcing decisions in an effort to meet a firm's long-range plans and goals. Strategic sourcing consists of strategic purchasing, internal integration, information sharing, and supplier development. It helps to select a group of strategic suppliers to develop a possible long-term partnership. "Internal integration enables purchasing to understand the needs of other functions like design, R&D and production" (Chiang, Kocabasoglu-Hillmer & Suresh, 2012, p. 69). Furthermore, the close relationship and the strategic role of purchasing, provide a foundation to conduct joint planning, response to market demand change, and satisfy specific customer requirements on the product (Chiang, Kocabasoglu-Hillmer & Suresh, 2012).

Purchase strategically requires a deep understanding of the assets traded on the needs of the buyer and supplier and also about the nature of their relationship (Menita et al., 2011). Sourcing professionals' activities concentrate on (1) supplier selection, which entails identifying suppliers and performing a comparative evaluation of suppliers' abilities to meet sourcing requirements, and (2) supplier governance, which involves designing and negotiating contracts as well as implementing mechanisms to interact with internal customers and suppliers, to ensure the successful fulfillment of the sourced product/service (Rai & Hornyak, 2013).

The skills required for procurement professionals became more evident after the intensification of international trade and the opening of global frontiers in the late 1990s. This opening exponentially expanded the network of possibilities for the supply and diversity of criteria that now permeate the decision-making process (Boer, Labro & Morlacchi, 2001). These factors demanded that the choice of suppliers assume a more strategic focus, which happened to be building more enduring relationships. These partnerships would reduce the number of suppliers able to be chosen by restricting the list to only the most reliable (Ho, Xu & Dey, 2010).

Companies that recognize the value of purchasing strategy have an area of proactive purchasing, with skills and resources necessary to carry out operations with strategic level (Carr & Smeltzer, 1997). Strategic sourcing requires a long-term orientation and may ultimately create a collaborative advantage and bring about greater benefits through collaborative advantage than a traditional nonstrategic sourcing-based approach to competition. Especially enhanced buyer-supplier relationships through information sharing and supplier development practices may be regarded as a competence and expected to improve performance and competitiveness (Chen, Paulraj & Lado, 2004; Chiang, Kocabasoglu-Hillmer & Suresh,

2012). So, it is imperative for practitioners to incorporate strategic sourcing as an integral part of the firm's business processes (Su, 2013).

2.2 Supplier-evaluation system

A supplier-evaluation system is defined as one whose activities are undertaken by the buying firms in their efforts to measure and improve the products or services they receive from their suppliers (Prahinski & Benton, 2004). The evaluation of suppliers is a tool used to gain advantage over competitors, and through this assessment information is obtained and will promote the development of joint processes and strategies that will guide shares between buyers and suppliers, increasing qualities and decreasing costs (Chow, Heaver & Henriksson, 1994). Supplier-evaluation process is a quantification process designed to stimulate the decision process inside the evaluating buying company or through the incentives it invokes, to stimulate a change in behavior in the evaluated supplying company (Neely et al., 1997). It is a connected entity, which broadens performance measurement analysis within supplier evaluation practices, extending it from a study of single contingencies to a study of an interrelated chain of actor interference, decision making, and communication (Hald & Ellegaard, 2011).

When companies outsource a significant part of their business, the processes of supplier evaluation become strategic. Today, it is important to understand how suppliers work: your business, your work process, your capabilities, and ultimately establish a relationship among companies (Liker & Choi, 2004). This requires formalization in the buyer-supplier relationship, with contracts to coordinate the relationship guarantees the rights of the companies involved and the legal borders of the activities of the development process (Sobrero & Roberts, 2001). For example, Carr and Pearson (1999) found that formal communication of supplier evaluations positively influenced the supplier-evaluation system.

Based on the evaluation process, the buying firm can determine if the supply base is capable of meeting current and future business needs. The buying firm needs to quantify and communicate the measurements and targets to the supplier, so that the supplier is made aware of the discrepancy between its current performance and the buying firm's expectations (Prahinski & Benton, 2004). It involves the creation mechanisms and procedures that ensure the exchange of information and knowledge between the parties during the development of products, and partnerships provide opportunities for learning during the process (Sobrero & Roberts, 2001).

When the buying firm uses collaborative communication for the supplier-development programs, it is perceived by the supplier as an effective mechanism to improve the buyer–supplier relationship. Collaborative communication includes indirect influence strategy, formality, and feedback. However, the implementation of several supplier-evaluation communication strategies by itself is not enough to influence the supplier's performance. Relationship development includes enhancing cooperation; problem solving; and expressing commitment, loyalty, and a desire to continue the relationship for many years into the future (Prahinski & Benton, 2004).

Effective evaluation of outsources' capabilities and relationship management are often central for outsourcers to secure sustainable competitive advantage. Zhang et al. (2012) investigate how to evaluate outsources and manage outsourcing relationships in the pharmaceutical industry based on the theory of dynamic capability. Their study shows that a company can successfully pursue both strategic and operational outsourcing simultaneously by applying different supplier evaluation criteria and relationship management methods. Hence, a company can source new external knowledge and resources and reduce operational costs at the same time, which provides a way to tackle the potential negative consequences associated with outsourcing.

There are numerous perceived benefits to the supplier evaluation system. Among those: (1) meet the suppliers in more detail, (2) correct procedures and practices that can help suppliers to obtain better performance, (3) based on a diagnosis, more specialized, forward future actions for the benefit of the best suppliers, (4) improve the supplier's opinion about their practices, previously grounded only in reducing cost, and (5) achieve improvements in different areas, including increase competitiveness and extend gains for the entire organization (Neumann & Ribeiro, 2004).

2.3 Buyer-supplier relationship

Companies that establish long-term relationships with key suppliers can move the company to have an improvement in financial performance (Watts & Hahn, 1993; Carr & Pearson, 1999; Cohen & Silva,

2000; Chen, Paulraj & Lado, 2004). However, the relationships are not rare or difficult to imitate. Byers can only achieve a differential advantage if they bring greater bargaining power to the table. It is the collaboration between firms that can generate relational rents though relation-specific assets, knowledge-sharing routines, complementary resource endowments, and effective governance. Collaborative advantage comes from relational rent, a common benefit that accrues to collaborative partners through combination, exchange, and co-development of idiosyncratic resources (Dyer and Singh, 1998).

The relational view takes the inter-organizational level of analysis and addresses the extent to which relational capabilities form the basis of durable strategic advantages (Dyer and Singh, 1998). Such a strategic intent then drives firms to acquire, access, or develop additional resources through cooperation. Paulraj, Lado, and Chen (2008) define inter-organizational communication as a relational capability, which functions as an important mediating construct that has different effects on outcomes for supplier and buyer firms. The author emphasizes that for supplier, the adoption of a long-term relationship orientation is necessary but not sufficient for achieving strategic advantage; it is need to hone skills for effective communication in order to reap fully the benefits of long-term relationships with buyer firms. For the buyer firms, establishing a network form of governance may not be sufficient for achieving a strategic advantage; such a governance form may only engender strategic advantage through providing an inter-organizational context that is conducive to collaborative communication.

Thus, a nuanced understanding of the roles of these factors in shaping an inter-organizational exchange context that is conducive to collaborative communication is key to effectively managing buyer-supplier relationships for mutual benefits (Chen, Paulraj & Lado, 2004; Paulraj, Lado & Chen, 2008). Collaboration with suppliers can provide elements of optimization and cost reduction; moreover, the dimension exchange of information or communication presents the relationship between buyer and supplier that can positively influence profitability (Carr & Pearson, 1999; Conceição & Quintão, 2004; Chen, Paulraj & Lado, 2004; Paulraj, Chen & Flynn, 2006).

Cao & Zhang (2011) identified a set of seven interconnecting dimensions that make up effective supply-chain collaboration: information sharing, goal congruence, decision synchronization, incentive alignment, resource sharing, collaborative communication, and joint knowledge creation. Authors confirmed that collaborative advantage and well-executed supply-chain collaboration directly improves firm performance in the long run. Collaborative advantage can be understood as a function of the combined value and rarity of all shared resources among supply-chain partners. The relationship implies that, in order for a supply chain as a whole to perform well, firms should try to create a win-win situation where all participants collaborate to achieve business synergy and compete with other chains (Paulraj, Chen & Flynn, 2006).

This relationship is beneficial for both sides, and the exchange of knowledge implies reduced spending for suppliers to reduce the time looking for faults and their causes. A good relationship is, for buyers, the capacity to share information of the products purchased (Watts & Hahn, 1993). The buying firm needs to establish an environment that is conducive to improving buyer–supplier relationship. Relationship development includes enhancing cooperation, problem solving, and expressing their commitment, loyalty, and desire to continue the relationship for many years into the future (Prahinski & Benton, 2004). Although, problem solving is not always regarded as something to value in a relationship, Brito, Brito & Hashiba (2014, p. 958) investigated the relationship between customers and suppliers and found that "shared problem solving is not an attractive practice in the relationship with suppliers and customers in the packaging industry". Monitoring the relationship is necessary; close social interactions between buyer-supplier makes the buyer able to gain access to valuable resources and exploit synergies created in the relationship. It promotes the risks of opportunism, loss of objectivity, ineffective decision making and higher cost (Villena, Revilla & Choi, 2011).

The firm needs to manage its supply chain and establish trust-based working relationships with suppliers; the results can be supply-chain capabilities or intangible resources that are so unique to that company that it gives them an advantage that ultimately increases firm performance (Paulraj, Lado & Chen, 2008). The underlying belief is that the elevated status of the purchasing function can promote collaborative relationships with suppliers through increased trust and commitment between internal customers and external suppliers (Paulraj, Chen & Flynn, 2006).

2.4 Financial Performance

Financial performance is perceived by organizations as a result of reaching their economic goals. Among the more traditional indicators, the following may be cited: revenue growth and sales, the number of new customers, markets and strategies, cost management, working capital, return on investment and productivity/efficiency (Venkatraman & Ramanujam, 1986). The performance evaluation has several financial criteria, such as return on investment (ROI), return on sales (ROS), return on assets (ROA), and increased sales and market share. The improvement of these indices that represent the financial performance of companies requires the constant development of strategies that optimize business management across the organization, especially in the supply area, so that the acquisition of raw materials to production processes become more strategic (Vickery *et al.* 2003; Menita *et al.*, 2011).

There are several approaches to the indicators that best represent the financial performance of companies. The correct choice of these indicators will allow partnerships to evaluate the combined performance and identify actions to be performed, based on collaborative relationships seeking to achieve goals of the chain and not individual goals (Aragão *et al.*, 2004). Conceição & Quintão (2004) verified, in order to ascertain whether collaboration with suppliers and buyers influencing the performance of soft drink manufacturers, that the effect of collaboration in performance manifests itself far more effective than financial performance in operational and general. These efforts focus on long-term rather than short-term relationships between buyers and suppliers; companies can help both buyers and supplier reduce their costs with the possibility of achieving a competitive advantage.

Carr and Pearson (1999) identified that strategic purchasing, supplier evaluation systems, and buyer-supplier relationships are in fact important with respect to a firm's financial performance. Chen, Paulraj and Lado (2004) operationalized financial performance for the buying firm by items, indicating the extent of changes in: (1) return on investment; (2) profits as a percent of sales; and (3) net income before tax over the past 3 years. The authors demonstrate robust support for the links between strategic purchasing, customer responsiveness, and financial performance of the buying firm. This demonstrates that purchasing contribution directly to the firm's bottom line is also a vitally important strategic part-

ner in fostering supply-management capabilities, which may generate durable strategic advantage.

Strategic purchasing can have a profound impact on supply chain performance (operational and financial indicators) for both buyer and supplier firms or relational, process, information, and cross-organizational team integration (Paulraj, Chen & Flynn, 2006). Furthermore, face-to-face planning and communication with key suppliers will benefit the buying firm in the long run. In addition, purchasing professionals perceive that suppliers are more responsive to their requirements when a cooperative type of relationship exists. All other things being equal, those firms that pursue cooperative-type relationships with key suppliers can anticipate some improvement in their firm's financial performance (Carr & Pearson, 1999).

3. RESEARCH METHODOLOGY

As reiterated above, the main objective of this study is to investigate which factors of buyer-supplier dyad relate to the financial performance of the manufacturing industry in Brazil. The design that characterizes this research is descriptive and correlational and explanatory, as 174 companies were studied to characterize the profile of the manufacturing industry in Brazil. The prospective study was cross-sectional, because variables were analyzed on a single point in time. The methodological approach was quantitative, through primary data collection, using a survey questionnaire with closed questions. According to Hair *et al.* (2005), quantitative research is an empirical investigation whose purpose is to outline and analyze phenomena, or evaluate programs to isolate key variables.

The target of this research was companies belonging to the manufacturing industry in Brazil that are considered important for the formation of the Brazilian GDP. In addition, these companies represent relevant segments in national economic potential and may be cited segments: automakers, foodstuff (exporters), cleaning and hygiene (multinational), pharmaceutical (MNCs), and other sectors.

The total population of the survey was comprised of approximately 1,200 companies in various sectors of national industry, such as food; automotive; glassware; textiles; pharmaceuticals; steel and metallurgical; hygiene and cleaning products; machinery and equipment; paper and pulp. These industries were selected because they are part of a catalog of business buyers of belts and synchronized pulleys, and imported material, represented in Brazil by a single supplier, located in Santa Catarina.

The survey was conducted from January 5, 2011, until April 23, 2012. Answers were collected through questionnaires sent via email directly to the respondents or delivered in personal visits during this period. The objective of this research is to raise the factors of buyer-supplier dyad related to financial performance, thus, the questionnaires were sent to people whose responsibility is to meet and negotiate with their suppliers. Many companies do not have a specific job to buyers; so, the questionnaire possessed a clear instruction about who was able to answer the questions.

In a total of two stages, 1,080 questionnaires were sent electronically. Emails to 120 firms were not delivered successfully. The return evidenced by the emails were 622 read; 36 unread, 238 not successfully delivered. There was also the need to be forwarded 66 emails a temporary absence, there was no other evidence of the procedure adopted in the company. At the end of the survey on April 23, 2012, we received responses to 218 emails and over 97 printed questionnaires. Of the total responses, 174

companies in the universe of 1,200 companies and in some of these companies were more than one respondent, totaling 312 respondents (valid questionnaires), following the same methodology used by the authors Carr and Pearson (1999). It is observed that 03 questionnaires have missing values, so they were removed of the total received (315).

Data were collected through a survey using a translated questionnaire with 16 closed questions on a 5-point Likert model, applied and validated Carr and Pearson (1999). The questionnaire, translated into Portuguese, was performed considering the necessary adaptations to the Brazilian market. Characteristics were considered and presented in the format of questions for analyzing perception of respondents. The validation was conducted with a pre-test: 44 questionnaires sent by emails in the month of January 2011 and 20 printed questionnaires, totaling 64 questionnaires. After the necessary adjustments with the vocabulary translated into Portuguese and endorsement of the respondents, the questionnaire was sent to companies. The constructs represented by their theoretical variables and are presented in Table 1. We replicate the Carr and Pearson model (1999) with the same parameters.

Table 1 – Dimensions of Relationship Buyer-Supplier

	Dimensions	Variables	Author
D1	Strategic sourcing	1. Long-term 2. Changes of the company's strategic plans 3. Relationships (commercial / financial /...) with suppliers	Carr and Pearson (1999)
D2	Supplier Evaluation System	4. Certification of suppliers 5. Performance monitoring of suppliers 6. Assessment and recognition of supplier performance	
D3	Buyer-supplier relationship	7. Special arrangements with suppliers 8. Loyalty to the main suppliers 9. Meetings or meetings with key suppliers 10. Direct communication between the top management / managers of the company and key suppliers 11. Exchange system data / information (EDI) with key suppliers 12. Influence on the company by the main suppliers	
D4	Perceived financial performance	13. Return on investment of the company due to the buyer-supplier relationship 14. Improvement in profits from sales due to buyer-supplier relationship 15. Improvement in gross profit due to buyer-supplier relationship 16. Changes in present value / PV company in the last five years	

Source: Carr and Pearson (1999)

3.1 Common method variance

The dependent variable was collected with the same instrument that was used for our independent variables, so the correlation between them could be an artifact of the method. Thus we followed the suggestion of Podsakoff *et al.* (2003) and Cheung, Myers & Mentzer (2010) and took several procedural measures to control common method bias, such as: (1) different scales were used for dependent and independent variables to reduce method bias caused by scale effect commonalities, (2) we assured respondents' anonymity and confidentiality, and (3) conducting a pre-test and having items reviewed by academians and experts to avoid item ambiguity. In addition to procedural controls, we assessed the presence of common method variance via Harman's one-factor test (Podsakoff & Organ, 1986) by performing a factor analysis on all items, and the test suggests that common method variance did not pose a significant problem.

3.2 Approach to data analysis

In the data analysis, we used SPSS ® version 19, for a descriptive and exploratory analysis. We used Confirmatory Factor Analysis (CFA) to establish dimensionality, validity, and reliability of construct measurement. After this phase, we used the technique of multiple linear regressions to answer the research objective.

Every method has limitations. According to Vergara (2000, p. 59), "It is healthy anticipate the criticism that the reader can do the work, explaining which limitations the chosen method offers, but still justify it as the most adequate for the purposes of research". Given this, considering the Brazilian context, we present as a limitation of the research the universe of respondents, who were employed in most of the tactical and operational level organizations, organization that do not always have effective access to financial statements, income statements, and rates of evolution of the financial companies in five years according to the four dimensions of this research.

4. DESCRIPTIVE DATA ANALYSIS

This section will present the characteristics of businesses, the respondents, and the results of the averages, standard deviations, and correlations of the constructs used.

Table 2 – Companies Data

	Frequency	Percentage
Operational areas of the companies surveyed		
North	6	3.5%
Northeast	8	4.4%
Midwest	3	1. 5%
Southeast	15	8.8%
South	41	24.0%
National	80	45.9%
More than one region	21	11.9%
Total	**174**	**100%**
Segment of the surveyed companies		
Alimentary	25	14.65%
Textile	14	8.08%
Pulp and Paper	13	7.58%
Steel and Metallurgy	12	7.07%
Automobile	10	5.56%
Machines Manufacturer	9	5.05%
Plastic	9	5.05%
Construction	9	5.05%
Cleaning and Hygiene	8	4.55%
Pharmaceutical	7	4.04%
Other segments	58	33.32%
Total	**174**	**100%**
Number of employees		
Up to 500	63	36.4%
500 to 1000	32	18.2%
1,000 to 5,000	49	28.2%
5,000 to 10,000	10	5.6%
Above 10,000	20	11.6%
Total	**174**	**100%**
Revenue		
Up to 50MI	35	20.2%
50MI to 100MI	31	17.7%
100MI to 500MI	35	20.2%
500MI to 1.000BI	35	20.2%
Above 1.000BI	25	14.1%
Uninformed	13	7.6%
Total	**174**	**100%**

Table 2 shows that most companies operate in the domestic market, with a rate of 45.9%. They are located predominantly in the southern region with 24%. And, 11.9% perform activities in more than one region. You can see that the north, northeast, and Midwest are those with the lowest number of established companies.

Companies surveyed obtained their segments evenly distributed, with emphasis on the food sector with 14.65%.

The number of employees and turnover variables were used to identify companies' size. It can be seen that 45.40% of the companies employ more than 1,000 employees, indicating that they are large, as classified by Sebrae (2013), which considers large companies as those with more than 500 employees. Classification of companies was determined by BNDES (National Development Bank) (2013), in which groups with revenues up to 100MI are classified as an "average" company and above 100MI are classified as a "medium-large" and "large" company, with the latter groups totaled 54.5 % of the companies surveyed. In Table 3 we present the profile of the respondents.

Table 3 - Respondents Data

	Frequency	Percentage
Sector		
Maintenance / Warehouse	98	31.4%
Shopping / Commercial	166	53.2%
Adm. / Financial	37	11.9%
Board of Directors	11	3.5%
Total	**312**	**100%**
Position		
Auxiliary	63	20.3%
Analyst	138	44.1%
Supervisor / Head chief	86	27.7%
Manager	22	7.1%
Director	3	1.0%
Total	**312**	**100%**
Company time		
Up 1 year	27	8.6%
1 to 5 years	115	36.7%
5 to 10 years	72	23.2%
Above 10 years	98	31.5%
Total	**312**	**100%**
Education		
Through high school	40	12.9%
Graduation	213	68.2%
MBA	59	19.0%
Total	**312**	**100%**

It is observed in Table 3 that the majority of respondents, a total of 53.2%, belong to the purchasing department. This result is expected, since this is one of the functions with greater contact with a supplier, followed by maintenance industry/warehouse, which in many companies represent the purchasing department, with 31.4%. They represent, together, 84.6% of total respondents.

On the issue of position, most of the respondents are in the sphere of auxiliary or analyst, totaling 64.4%. The questionnaire was sent to the head of the purchasing department, thus, this result indicates that while, even with this being a strategic sector for companies, they have not yet awakened to their real importance in the performance of their business.

Also according to Table 3.54.7% of respondents have over 5 years of work experience and 36.7% have between 1 and 5 years work experience. These results are significant, because the more time working for a company, the more a respondent can understand their activity and also the company it serves. In addition, it was found that 68.2% of the respondents have a higher education, and 19% of the respondents have a MBA, indicating that they are qualified for the sector that acts with the appropriate level for the understanding of the questions referred.

The scales presented in this study were validated and confirmed by Carr and Pearson (1999) and adequately represent their constructs. In Table 4 it is presented the correlation matrix and descriptive data of all sizes used.

Table 4 - Means, standard deviations, and correlations

	Mean	SD	Strategic sourcing	Supplier Evaluation System	Buyer-supplier relationship	Perceived financial performance
Strategic sourcing	4.0	0.98	-			
Supplier Evaluation System	3.75	1.15	0.548**	-		
Buyer-supplier relationship	3.81	0.76	0.519**	0.474**	-	
Perceived financial performance	3.77	0.89	0.450**	0.355**	0.664**	-

$*(p<0.10)$; $**(p<0.05)$; $***(p<0.01)$

Table 4 shows the results are within the normal relationships between the dimensions worked with significance level <0.01.

4.1 Reliability of dimensions

All dimensions of the Carr and Pearson (1999) model were perceived by survey respondents. A three-stage continuous improvement cycle was used to develop measures that satisfied all the requirements for reliability, validity, and unidimensionality (Chen & Paulraj, 2004). To assess the reliability of the study constructs, we used the average correlation among items in a scale. The Cronbach's alpha values for

the variables were well above 0.70 (Hair et al., 2005). Strategic sourcing (0.835), supplier-evaluation system (0.763), buyer-supplier relationship (0.899), and financial performance (0.736).

Confirmatory factor analysis (CFA) was used to assess construct validity and unidimensionality. CFA provides a stricter and more-precise test of unidimensionality of latent constructs. From the original model fit, absolute fit measures were employed in selected cases, such as chi-square likelihood ratio ($\chi 2$) and root mean squared residue, in order to ensure adequate representation of the entire set of relations dimensions causal shown in Table 5.

Table 5: Chi-square

Chi-square	89.990
Degrees of freedom	39
Probability level	0.000
χ^2/GL	2.307
RMSEA	0.065

A table 5 show what is sought is a value not significant chi-square, since H0 indicates that data fit the model. In case of significant, Hair *et al.* (2005) argues that we can divide the value by the degrees of freedom (χ^2/GL), with 5 or less acceptable values. Complementing the chi-square, the root square error of approximation (RMSEA), which resulted in a value of 0.065, representing quality in model fit was performed, as recommended values vary between 0.05 and 0.08 (Kline, 2005; Brown, 2006, p. 87).

In addition to the measures presented, other measures of incremental adjustment were conducted: Normed Fit Index – NFI (0.960), Tucker-Lewis Index or Non-Normed Fit Index – NNFI (0.967) and Comparative Fit Index – CFI (0.977). Levels of quality adjustments are appropriate, with values above 0.90 (Kline, 2005; Brown, 2006, p. 87).

The model showed levels of reliability and discriminate validity, indicating that each construct is unidimensional. A significant statistical difference chi-square (χ^2) for the two models aligned value in-

dicates that the constructs are different and assures discriminate validity (O'leary-Kelly & Vokurka, 1998). Testing of all pairs of constructs was performed, showing a difference in the statistical chi-square (χ^2) significantly (p <0.05).

Finally, we have observed the multicollinearity of dimensions with VIF test, to verify if dimensions can be used without any further action. The results showed, D1 - strategic sourcing (1.953), D2 - supplier Evaluation System (1.956), D3 - buyer-supplier relationship (1.648). The standard way to assess the magnitude of the multicollinearity problem is the variance inflation factor (VIF) scores for the variables in each regression model. VIF scores above 10 indicate a serious problem (Cohen *et al.* 2002). VIF scores are below 5 most below 3. These results indicate that multicollinearity is not a major concern, because all VIF scores are below 2.

4.2 Regression analysis

Regression was crafted considering the perceived financial performance scale as the dependent variable and the dimensions strategic sourcing, supplier evaluation system and Buyer-supplier relationship as independent.

The model shows up with significant determination coefficient in that the dimensions strategic sourcing supplier evaluation system Buyer-supplier relationship explained 45.5% (table 6) of the construct financial performance.

Table 6 – Coefficients

	Beta	Std. Error	t	Sig
(Constant)	1.481	1.159	1.278	0.202
Companies Control Variables				
Industrial sector				
Alimentary	-0.175	0.141	-1.244	0.215
Automobile	-0.518	0.217	-2.390	0.018**
Construction	-0.099	0.303	-0.326	0.745
Pharmaceutical	0.018	0.256	0.069	0.945
Cleaning and Hygiene	0.115	0.213	0.541	0.589
Machines Manufacturer	0.467	0.239	1.958	0.051*
Pulp and Paper	-0.141	0.173	-0.816	0.415
Plastic	-0.497	0.264	-1.882	0.061*
Steel and Metallurgy	-0.257	0.180	-1.423	0.156
Textile	-0.119	0.163	-0.728	0.467
Other segments	-0.213	0.759	-0.281	0.779
Firm Size - Revenue				
Up to 50MI	-0.033	0.213	-0.155	0.877
50MI to 100MI	0.139	0.196	0.706	0.481
100MI to 500MI	0.078	0.195	0.401	0.689
500MI to 1.000BI	0.017	0.195	0.087	0.931
Above 1.000BI	0.043	0.205	0.210	0.834
Dimensions				
D1 - Strategic sourcing	0.228	0.055	4.166	0.000***
D2 - Supplier Evaluation System	0.022	0.048	0.463	0.643
D3 - Buyer-supplier relationship	0.473	0.065	7.309	0.000***
R 0.674				
R^2 0.455				
R^2 (adjusted) 0.367				

*($p<0.10$); ** ($p<0.05$); *** ($p<0.01$)

We haven't found significant statistic relationship on respondents control variables showing that the financial performance perception is not related with sector where they work, position, time with the company, and education. However, we have found significant statistic relationship on companies' control variables of segment, showing that automobile companies have a lower perception of financial performance, and machine manufacturers and plastic companies have a higher perception of financial performance. The other sectors as well as other companies' control variables do not indicate any significant statistic relationship with financial performance perception.

Finally, the coefficients indicate a positive relationship between the dimensions – strategic sourcing and buyer-supplier relationships - with the company's financial performance dimension. However, the supplier evaluation system showed no statistically significant relationship with a Sig quite high.

Strategic sourcing requires an understanding of the needs of both buyer and supplier. It takes an understanding of the nature of their relationship (Menita *et al.*, 2011). Companies that recognize the value of purchasing strategy have an area of proactive purchasing, with skills and resources necessary to carry out operations with strategic level (Carr & Smeltzer, 1997). Furthermore, such companies conduct joint planning, respond to market demand change, and satisfy specific customer requirements on the product (Chiang, Kocabasoglu-Hillmer & Suresh, 2012). Sourcing professionals' activities concentrate on supplier selection and supplier governance (Rai & Hornyak, 2013).

Strategic sourcing requires a long-term orientation and may ultimately create collaborative advantage and bring about greater benefits of collaborative advantage than a traditional non-strategic source-based approach to competition (Chen, Paulraj & Lado, 2004; Chiang, Kocabasoglu-Hillmer & Suresh, 2012). When planning, the buyer sector increases its strategic role, enabling the development of long-term negotiations, which consequently generates relationships that may encourage the development of suppliers through innovation, improved product quality, and reduced costs, among other factors (Conceição & Quintão, 2004; Lima, 2008), promoting value creation in the relationship favoring both sides (Corsten & Kumar, 2005).

The role of purchasing passes from transactional to relational, and this new structure affects the positioning of the sector within the organization; aligned with the company's strategic planning, purchasing actions reflect on its performance. The relational view is extended to relational capabilities form the basis of durable strategic advantages (Dyer and Singh, 1998).

The relationship implies that, in order for a supply chain as a whole to perform well, firms should try to create collaboration to achieve business synergy and compete with other chains (Paulraj, Chen & Flynn, 2006). Another key factor is collaborative communication to effectively manage buyer–supplier relationships for mutual benefits (Chen, Paulraj & Lado, 2004; Paulraj, Lado & Chen, 2008; Cao & Zhang, 2011). Relationship development includes enhancing cooperation, problem solving, and expressing the commitment, loyalty, and desire to continue the relationship for many years into the future (Prahinski & Benton, 2004).

Collaboration with suppliers can provide elements of optimization and cost reduction (Carr & Pearson, 1999; Conceição & Quintão, 2004; Chen, Paulraj & Lado, 2004; Paulraj, Chen & Flynn, 2006). It is important that the purchasing department can demonstrate how they, through the buyer-supplier relationship, may increase the company's financial performance. Companies that prioritize this relationship, according to Cohen and Silva (2000), perceive improvements in financial return on their investment and indexes higher financial performance (Carr & Pearson, 1999).

However, collaborative communication includes indirect influence strategy, formality, and feedback. This formalization in the buyer-supplier relationship involved legal borders of the activities of the development process (Sobrero & Roberts, 2001). Supplier evaluation system can be improved the products or services they receive from their suppliers (Neely *et al.*, 1997; Prahinski & Benton, 2004). But for this, the buying firm needs to quantify and communicate the measurements of the discrepancy between its current performance and the buying firm's expectations (Prahinski & Benton, 2004).

The lack of formalization in the buyer-supplier relationship, contracts, or evaluation system for the coordination of the relationship as well as mechanisms and procedures to ensure the exchange of information on certifications, supplier performance, and product development and partnerships, make it difficult to measure how important is the supplier for company (Sobrero & Roberts, 2001).

In contrast with results found by Carr and Pearson (1999), in which formal communication of supplier evaluations positively influenced supplier evaluation system, the results of this regression show that formalization and evaluation systems suppliers are still in the early stages. This shows how much the buyer-supplier relationships are informal in the manufacturing industry in Brazil. This fact generates insecurity in the commitment of the relationship, negatively influencing knowledge exchange, shared learning, and confidence in the relationship of dyad. The other results of the survey were consistent with Carr and Pearson (1999).

5. CONCLUSION AND FURTHER RESEARCH

Our study contributes to and continues a growing research stream about dyad buyer-supplier and its relation to financial performance. Specifically, it investigates the relationships among strategic sourcing, supplier-evaluation system, buyer-supplier relationship, and perceived financial performance.

The objective of this study was to raise which practices of buyer-supplier dyad are related to the financial performance of the manufacturing industry in Brazil. The results showed a positive relationship between the dimensions--strategic sourcing and buyer-supplier relationships--with the company's financial performance dimension.

The procurement planning considers the existence of formal planning long-term, if it is reviewed regularly and includes various forms of relationship between buyer and supplier. Through the perception of the respondents, the Brazilian manufacturing industry demonstrates a high level of planning in purchasing. In the analysis of the relationship between buyer and supplier, the data showed that the special agreements with suppliers and fidelity are considered important factors, followed by the influence of the largest suppliers and communication between senior management and key suppliers.

The positive outcome of these two dimensions--strategic sourcing and buyer-supplier relationships--with the company's financial performance is justified when considering the increasing reliance of businesses upon their suppliers, and therefore the need for more planning and better relationships near the dyad. However, this study also showed that the supplier-evaluation system showed no statistically significant relationship with financial performance. This result reinforces the lack of formalization in the buyer-supplier relationship, a situation that often hinders the development of a long-term relationship. The lack of systematic evaluations of supplier performance can generate insecurity in the relationship, since historical actions taken by suppliers could serve as a criterion of choice in a future negotiation.

As practical contributed to this study, it is expected that managers understand the importance of buyer-supplier relationship to the company's financial performance. More specifically, it is important that they perceive the importance of evaluating their suppliers. The evaluation of suppliers is an important practice to qualify and improve.

For further research, it appears that other sectors should be studied, and they should be analyzed by the buyer-supplier dyad; construct financial performance could be measured based on financial data and non-perception as used in this study, and the relationship of control variables (automobile, machines manufacturer, and plastic) with financial performance could be worked on a qualitative view.

REFERENCES

Anderson, M. G., & Katz, P. B. (1998). Strategic sourcing. *International Journal of Logistics Management, 9*(1), 1.

Aragão, A. B., Scavarda, L. F., Hamacher, S., Pires, S. R. I. (2004). Modelo de Análise de Cadeias de Suprimentos: Fundamentos e Aplicação às Cadeias de Cilindros de GNV. *Gestão & Produção, 11*(3), 299-311.

BNDES. Classificação das empresas. Avaliable on http://www.bndes.gov.br/SiteBNDES/bndes/bndes_pt/Institucional/acesso_a_informacao/. Access on 09/02/2013.

Brito, L. A. L., Brito, E. P. Z., Hashiba, L. H. (2014). What type of cooperation with suppliers and customers leads to superior performance? *Journal of Business Research, 67*, 952-959.

Boer, L.; Labro, E., & Morlacchi, P. (2001). A Review of Methods Supporting Supplier Selection. *European Journal of Purchasing and Supply Management, 7*(2), 75-89.

Brown, T. (2006) *Confirmatory factor analysis for applied research.* New York: The Guilford Press.

Cao, M., & Zhang, Q. (2011). Supply chain collaboration: Impact on collaborative advantage and firm performance. *Journal of Operations Management, 29*(3), 163–180.

Carr, A. S., & Pearson, J. N. (1999). Strategically managed buyer–supplier relationships and performance outcomes. *Journal of Operations Management, 17*, 497–519.

Carr, A. S., & Pearson, J. N. (2002). The impact of purchasing and supplier involvement on strategic purchasing and its impact on firm's performance. International. *Journal of Operations & Production Management, 22*(9), 1032-53.

Carr, A. S., & Smeltzer, L. R. (1997). An empirically based operational definition of strategic purchasing. *European Journal of Purchasing and Supply Management, 3*(4), 199–207.

Chen, I. J., Paulraj, A., & Lado, A. A. (2004). Strategic purchasing, supply management, and firm performance. *Journal of Operations Management, 22*(5), 505-23.

Chen, I.J., Paulraj, A. (2004). Towards a theory of supply-chain management: the constructs and measurement. *Journal of Operations Management, 22*(2), 119–150.

Cheung, M., Myers, M. B., Mentzer, J. T. (2010). Does relationship learning lead to relationship value? A cross-national supply chain investigation. *Journal of Operations Management, 28*, 472–487.

Chiang, C.-Y., Kocabasoglu-Hillmer, C. & Suresh, N. (2012). An empirical investigation of the impact of strategic sourcing and flexibility on firm's supply chain agility. *International Journal of Operations & Production Management, 32*(1), 49-78.

Cohen, J., Cohen, P., West, S. G., & Aiken, L. S. (2003). Applied multiple regression/correlation analysis for the behavioral sciences (3rd ed.). Mahwah, NJ: Lawrence

Erlbaum Associates.

Chow, G., Heaver, T., & Henriksson, L. (1994). Logistics Performance: Definition and Measurement. *International Journal of Physical Distribution & Logistics Management*, 24(1), 17-28.

Cohen, M., & Silva, J. F. (2000). O Impacto das decisões estratégicas no desempenho dos franqueados em fast-food: o papel do relacionamento franqueador-franqueado. *Revista de Administração Contemporânea – RAC*, 4(2).

Conceição, S. V., & Quintão, R. T. (2004). Avaliação do Desempenho Logístico da Cadeia Brasileira de Suprimentos de Refrigerantes. *Gestão & Produção*, 11(3), 441-453.

Corsten, D., & Kumar, N. (2005). Do Suppliers Benefit from collaborative Relationships with Large Retailers? An Empirical Investigation of Efficient Consumer Response Adoption. *Journal of Marketing*, 69, 80–94.

Dyer, J., & Singh, H. (1998). The relational view: Cooperative strategy and sources of interorganizational competitive advantage. *The Academy of Management Review*, 23(4), 660–679.

Ellram, L. M., & Liu, B., 2002. The financial impact of supply management. *Supply Chain Management Review* 6(6), 30–37.

Fiesp. (2014). Panorama da indústria de transformação brasileira. Avaliable on < file:///C:/Users/User/Downloads/panorama-da-industria_3a-edicao%20(2).pdf>. Acess on 09/06/2014.

Hair Jr., J. F., Babin, B., Money, A. H., & Samouel, P. (2005). *Fundamentos e métodos de pesquisa em administração*. Porto Alegre: Bookman.

Hald, K. S, & Ellegaard, C. (2011). Supplier evaluation processes: the shaping and reshaping of supplier performance. *International Journal of Operations & Production Management*, 31(8), 888-910.

Ho, W., Xu, X., & Dey, P. K. (2010). Multi-criteria Decision Making Approaches for Supplier Evaluation and Selection: A Literature Review. *European Journal of Operational Research*, 202(1), 16-24.

Joshi, A. W. (2009). Continuous Supplier Performance Improvement: Effects of Collaborative Communication and Control. *Journal of Marketing*, 73, 133–150.

Kline, R. (2005). *Principles and practice of structural equation modeling*. 2. ed. New York: Guilford Press.

Lima, J. C. S. (2008) A configuração da área de compras e o trabalho do comprador nas montadoras: novas atribuições e perfil. *Gestão da Produção, Operações e Sistemas - GEPROS*, 3(1), 11-22.

Liker, J. K., & Choi, T. Y. (2004). Building Deep Supplier Relationships. *Harvard Business Review*, 82(12), 104-113, december.

Menita, P. R., Vanalle, R. M., Salles, J. A. A., & Oliveira, R. D. (2011) Análise das estruturas de governança como instrumento de gestão de compras estratégicas. *Revista Científica Indexada Linkania Master*, 1(1), set./out.

Narasimhan, R., & Das, A. (2001). The Impact of Purchasing Integration and Practices on Manufacturing Performance. *Journal of Operations Management*, 19(5), 593-609.

Neely, A. D., Richards, H., Mills, J., Platts, K. & Bourne, M. (1997). Designing performance measures: a structured approach. *International Journal of Operations & Production Management*, 17(11), 1131-52.

Neumann, C. S. R., & Ribeiro, J. L. D. (2004). Desenvolvimento de fornecedores: um estudo de caso utilizando a troca rápida de ferramentas. *Revista Produção*, 14(1).

O'leary-kelly, S. W, Vokurka, R. J. (1998). The empirical assessment of construct validity. *Journal of Operations Management*, 16(4), 387-405.

Paulraj, A., Lado, A., & Chen, I. J. (2008). Inter-organizational communication as a relational competency: Antecedents and performance outcomes in collaborative buyer-supplier relationships. *Journal of Operations Management* 26(1), 45-64.

Paulraj, A., Chen, I. J., & Flynn, J. (2006). Levels of strategic purchasing: Impact on supply integration and performance. *Journal of Purchasing & Supply Management*, 12(3), 107-122.

Pearson, J. N., Ellram, L. M., & Carter, C. (1996). Status and recognition of the purchasing function in the electronic industry. *International Journal of Purchasing and Materials Management*, 32(2), 30–36.

Porter, M. E. (1986). *Estratégia competitiva: técnicas para análise de indústrias e da concorrência*. Rio de Janeiro: Campus.

Podsakoff, P., Mackenzie, S., Lee, J., & Podsakoff, N. (2003). Common method biases in behavioral research: A critical review of the literature and recommended remedies.*Journal of Applied Psychology*, 88(5), 879–903.

Podsakoff, P., & Organ, D. (1986). Self-reporting in organizational research: Problems and prospects. *Journal of Management*, 12(4), 531–544.

Prahinski, C., & Benton, W. C. (2004). Supplier evaluations: communication strategies to improve supplier performance. *Journal of Operations Management*, 22(1), 39-62.

Rai, A., & Hornyak, R. (2013). The impact of sourcing enterprise system use and work process interdependence on sourcing professionals' job outcomes. *Journal of Operations Management*. 31(6),474–488.

SEBRAE. (2013). Classificação de empresas. Avaliable on <http://www.sebrae-sc.com.br/leis/default.asp?vcdtexto=4154.>. Access on 02/27/2013.

Sobrero, M., & Roberts, E. B. (2001). The trade-off between efficiency and learning in interorganizational relationships for product development. *Management Science*, 47(4).

Su, J. (2013).Strategic sourcing in the textile and apparel industry. *Industrial Management & Data Systems*, 113(1), 23-38.

Terpend, R., Tyler, B. B., Krause, D. R., & Handfield, R. B. (2008). Buyer–supplier relationships: derived value over two decades. *Journal of Supply Chain Management*, 44(2), 28–55.

Vickery, S.K., J. Jayaram, C. Droge and R. Calantone. (2003). The Effects of an Integrative Supply Chain Strategy on Customer Service and Financial Performance: An Analysis of Direct Versus Indirect Relationships. *Journal of Operations Management, 21*(5), 523-539.

Villena, V. H., Revilla, E., Choi, T. Y. (2011). The dark side of buyer–supplier relationships: A social capital perspective. *Journal of Operations Management, 29*(6), 561-576.

Venkatraman, N., & Ramanujam, V. (1986). Measurement of Business Performance in Strategy Research: A Comparison of Approaches. *Academy of Management Review, 11*(4), 801-814.

Vergara, S. C. (2000). *Projetos e relatórios de pesquisa em administração.* 3. ed. São Paulo: Atlas.

Watts, C., & Hahn, C. (1993). Supplier development programs: an empirical analysis. *International Journal of Purchasing and Materials Management, 29*(2), 10–17.

Zhang, M., Pawar, K. S., Shah, J. & Mehta, P. (2012). Evaluating outsourcing partners' capability: a case study from the pharmaceutical supply chain. *Journal of Manufacturing Technology Management, 24*(8), 1080-1101.

AUTHOR'S BIOGRAPHY

Luis Cesar Mondini - Graduated in Economics from the Regional University of Blumenau and Logistics from the University Center Leonardo DaVinci. Specialists in Business Logistics and Distribution from Instituto Catarinense de Pós-Graduação. Master in Business Administration from Furb. Professor at the University Center Leonardo Da Vinci - Uniasselvi.

Marcia Regina Santiago Scarpin - PhD student at FGV/EAESP (São Paulo, Brazil). Master in Business Administration from Furb. Professor of logistic and supply chain at the Centro Universitário Curitiba (Unicuritiba).

Denise Del Prá Netto Machado - PhD from the FGV/EAESP and Postdoctoral from the UnB. Professor at Regional University of Blumenau. Member of the Scientific Committee of the Area of Management Science, Technology and Innovation of ANPAD. Experience with emphasis in Organizational Culture, Innovation Management and Organizational Studies.

6

The Application of Lean Principles in the Fast Moving Consumer Goods (FMCG) Industry

Alaa Aljunaidi
alaajuna@gmail.com
Pepsi Cola - Bugshan (SIBCO)

Samuel Ankrah
samuel.ankrah@my.ohecampus.com
Specialist Health Ghana

ABSTRACT: Proponents of Lean Manufacturing (*lean*) claim that the benefits of implementing this system are not restricted to the automotive industry where lean originated. On the other hand, opponents of *lean* argue against its claimed universality and stress that even within the automotive industry mass production still prevails. This research aimed at examining this claimed universality by investigating whether *lean* is applicable to the Fast Moving Consumer Goods (FMCG) Industry. The research was carried out via qualitative, multiple case-studies of three different Saudi Arabia FMCG operations including two confirmatory interviews, one with a US FMCG manufacturer and another with a UK manufacturer. Data was collected via semi-structured interviews with 22 participants.

The research concluded that the same types of waste identified by *lean* in the automotive industry are present in the FMCG industry, suggesting that *lean* could be implemented in FMCG operations. The study also identified the conditions required for successful implementation and consequently developed a conceptual framework for implementing lean in the FMCG industry.

1. INTRODUCTION

In the 1980s it became apparent that the American auto industry which depended on mass production was losing to foreign competition. Consequently, a commission led by the Massachusetts Institute of Technology (MIT) was set up in 1986 to investigate the causes of the phenomenon (Duguay, Landry & Pasin, 1997; Womcak & Jones, 1996). The results of the study were published in The Machine That Changed the World (Womack, Jones & Roos, 1990, pp. 9-282), a book which documented the superior performance of Japanese Auto manufacturers due to their implementation of lean manufacturing (Womack & Jones, 1996). Womack & Jones (2003, p. 15) argue that Lean Manufacturing (lean) is the "antidote" to muda, the Japanese word for waste. In a way, lean is similar to mass production in the sense that it "creates outputs using less of every input" (Lehtinnen & Torkko, 2005, p. 58). However, the difference between the two lies in lean's ability to offer more variety to customers (Womack, Jones & Roos, 1990, pp. 11-12).

Since the early 1990s, lean has been successfully implemented across several industries, including appliance manufacturing, the aerospace industry, healthcare and fast food companies (Moyano-Fuentes & Sacristan-Diaz, 2012). However, notwithstanding, there is a fierce debate in the lean literature on the universality of lean, i.e. its applicability across all types of industries, including non-manufacturing (or service) industries. Whereas some authors like James-Moore & Gibbons (1997) agree with the claim that lean is a universal model applicable to any industry, others like Cooney (2002) disagree.

The literature reveals that not much attention has been given to the application of lean principles in the Fast Moving Consumer Goods industries (FMCG). The few articles on lean in FMCG appear to focus on either the supplier's end of the supply chain (for example, Found & Rich, 2007), or the retailer's end of the supply chain (for example, Francis, 2004), but not the operations of the FMCG manufacturing. This gap is clearly evident in the study carried out by Bhamu & Sangwan (2014) in which they reviewed 209 lean related literature published between 1988 and 2012. Of these, only one research (0.47%) was related to FMCG industry, and even this single research was related to the retail side of the business rather than the manufacturing side. This research therefore contributes to filling this gap by investigating the applicability of lean in FMCG manufacturing

and warehousing operations through case studies of three FMCG manufacturers located in Saudi Arabia, including two confirmatory interviews, one with a US FMCG manufacturer and another with a UK manufacturer.

Although lean is a well-established system that has been implemented for decades across several industries, it does not appear to have an agreed upon approach for implementation. Various initiatives in the lean literature have proposed implementation approaches, which could be grouped into five categories: conceptual frameworks, implementation frameworks, roadmaps, descriptive and assessment checklist (Mustafa, Dumrak & Sultan, 2013). This research proposes a conceptual framework specifically designed for lean implementation in the FMCG manufacturing industry.

The research question is: How can lean principles be applied in the FMCG manufacturing and warehousing operations? This research question is addressed through the following research objectives:

- To determine whether lean principles can be applied in FMCG operations.

- To examine the conditions under which lean principles can be applied in FMCG operations.

- To identify which lean tools and techniques can be applied in FMCG operations.

- To determine the potential benefits and drawbacks of implementing lean in FMCG operations.

We begin our inquiry by a review of lean literature on what lean is, the universality of lean and a quick review of lean tools and techniques. Next, we describe the research methodology. This is followed by a review of the research results. We then analyze and discuss the results and propose a conceptual framework for implementing lean in the FMCG industry. Through this analysis, we contribute to the lean literature by covering the gap which we found in the literature related to implementing lean in the FMCG industry, and on the basis of our study's outcome that lean is applicable in FMCG manufacturing and warehousing operations, also support the view that lean is a universal system.

2. LITERATURE REVIEW

2.1 The meaning of lean

The Machine That Changed the World is a book about the practices of the Japanese auto industry in general, and the Toyota Production System (TPS) in particular. *Lean* is the western name given to TPS. A clear-cut definition of *lean* is neither provided in *The Machine That Changed the World* (Womack, Jones & Roos, 1990, pp. 47-69) nor in the follow-up book, *Lean Thinking* (Womack & Jones, 2003, pp. 9-98). What is provided, however, are processes, tools and techniques to "banish waste and create wealth" (Womack & Jones, 2003, pp. 9-98) for any organisation.

Muda is the Japanese terminology for 'waste' (Womack & Jones, 2003, p. 15), and can be defined as "any human activity that absorbs resources but creates no value" (Wood, 2004, p. 8). Value, on the other hand, is the opposite of waste and is created as wasteful activities are reduced, or where additional features or services are added and are perceived to be valuable by the customer (Hines, Holweg & Rich, 2004). Taiichi Ohno, the Toyota executive who masterminded TPS identified seven types of waste (Womack & Jones, 2003, p. 15). Table 1 summarizes these wastes.

Table 1 - The Seven Types of Waste

2.2

Type of Waste	Definition	References
Overproduction	Producing goods without having customer orders, which results in overstaffing and higher costs of storage and inventory. This is the "most serious" of the seven types of waste as it obstructs the smooth flow of product or information and inhibits quality and productivity.	Liker (2004), p. 28, Hines & Rich (1997)
Waiting	Shop floor workers who sit idle either due to wrong workload balancing, just waiting for the next processing step, or because of out of stock of raw materials	Liker (2004), p. 28
Unnecessary Transport	The physical transportation of raw materials, finished goods or work-in-process (WIP). In addition to wasting time, this type of *muda* can also result in damage or deterioration of quality.	Liker (2004), p. 28, Hines & Rich (1997)
Over-processing	Performing unnecessary processing due to poor product or process design. Over-processing results in further muda in the form of defects, unnecessary movement and poor communication.	Liker (2004), p. 29, Hines & Rich (1997)
Excess Inventory	Inventory of raw materials, WIP or finished goods is a common way to hide problems (due to poor performance of suppliers or machines) and results in a significant increase in storage & transportation costs.	Liker (2004), p. 29, Hines & Rich (1997)
Unnecessary Movement	Any avoidable movement employees perform during their work which leads to poor productivity and possibly quality issues.	Liker (2004), p. 29, Hines & Rich (1997)
Defect	The production of defective goods. In the TPS, each defect is seen as an opportunity for improvement rather than a problem.	Liker (2004), p. 29, Hines & Rich (1997)

2.2 The universality of lean

Ohno (1988, p. 9), the creator of TPS, claims that the concepts of TPS, on which *lean* is based, "will work for any type of business." *Lean Thinking* (Womack & Jones, 2003, pp. 102-245), the follow up book to *The Machine That Changed the World*, is primarily a book

about success stories of *lean* implementations across various industries in various countries. However, there are critics of *lean* who debate this claimed universality, like Papahristodoulou (1994) who is of the view that *The Machine That Changed the World* is "a book that promises too much and proves too little." More recently, Howison (2009) has also claimed that:

"managerial fervour that described lean production as a lifeline for the US auto industry has subsided in the latest wave of bailouts and loans for General Motors and Chrysler. In spite of the adoption of lean production and the latest technological manufacturing equipment, 825 jobs were cut at Toledo's North Jeep Assembly Plant in October of 2008."

To such arguments and claims, the proponents of *lean* respond that whilst many companies claim that they are *lean*, they are in fact only implementing a set of *lean tools* without considering *lean* as a complete system (Liker, 2004, p. 10), for example, although just-in-time deliveries are claimed as being practiced, they are simply a relocation of inventories from a company to its next upstream supplier (Womack & Jones, 1996). The studies by Hines, Holweg & Rich (2004), Liker (2004), Spear & Bowen (1999) and Towill (2007) argue that *lean* is not only a

set of best practice tools that can be implemented to achieve results but rather a complete model, a philosophy that requires a change in culture in parallel with the change of work processes and procedures. According to Liker (2004, p. 10), this explains why many companies think they are *lean*, but are not in reality, because they do not implement the complete model, and this also explains why many *lean* implementations fail. It is, therefore, not surprising that Childerhouse & Towill (2004) found that the overall success rate of *lean* is as low as 50%.

2.3 Lean tools and techniques

With regard to the operational aspect of *lean* tools and techniques, Ohno (1988, pp. 4-15) stated that the two pillars of TPS are Just-In-Time (JIT) and Autonomation (or automation with a human touch), and these pillars are supported by teamwork and production levelling. Table 2 categorises the various *lean* tools and techniques into five groups

Table 2 - Lean tools

Category	Tool	Description	References
Just-In-Time (JIT) Tools	Kanban	Only produce to order, no more and no less, in order to create flow	Jones, Hines & Rich (1997)
	Andon	Giving operators the power to stop the operation when an error is detected	Jones, Hines & Rich (1997)
Autonomation tools	Jidoka	This means automated stop system and is a function of machine design to allow machine to stop when an abnormal situation arises	Ohno, T. (1988)
Production levelling tools	Heijunka	Production levelling by reducing batch size in order to make the best utilization of resources	Pettersen (2009)
	Quick changeovers	The ability to changeover from one product to another in a short period of time to allow small but more frequent production runs of the same product	Ohno (1988), pp. 95-97
	Takt-time	Takt is a German word which means beat or rhythm. This tool is used to synchronise the production rate (or rhythm) with customer demand in order to avoid Overproduction. The term originated in the German aircraft industry in the 1930s before it was later taken to Japan by German engineers training Japanese aircraft producers.	Simons & Zokae (2005)

Category	Tool	Description	References
Process improvement tools	*Value Stream Mapping Techniques*	After identifying value from the customer's perspective, the next step is to identify the value stream for each product	Hines & Rich (1997)
	Kaizen	This refers to continuous improvement events: One of the most basic yet powerful tools of lean. Once a problem is identified, staff from various functions and levels work together in order to find solutions to the problem	Chen, Li & Shady (2010)
	Kaikaku	Unlike Kaizen, these events involve major changes in the workplace including factory layout and machine rightsizing. They are typically carried out at the onset of any lean project	Womack & Jones (1996)
Visualisation tools	*Poka Yoke*	Failure prevention by designing machines in an 'error proof' way	Pettersen (2009)
	5S	A visualisation tool based on the conviction that organizing the workplace is essential to ensuring a smooth workflow. It is based on 5 Japanese words starting with the letter 's' which were later translated to equivalent English words: Seiri (Sort), Seiton (Set in Order), Seiso (Shine), Seiketsu (Standardise) and Shitsuke (Sustain).	Ravikumar, Marimuthu & Chandramohan (2009)

2.4 The FMCG industry

The FMCG industry makes a significant contribution to the world economy. For instance, in the UK, the industry employs over 16% of the total workforce and contributes over 8% of the UK's Gross Domestic Product (GDP) (Bourlakis & Weightman, 2004 cited in Francis, Dorrington & Hines, 2006). It is also the largest sector in New Zealand and the fourth largest in India (Economy Watch, 2010). In Saudi Arabia, although the country's economy is heavily dependent on the oil industry which represents 60% of total investments in the industrial sector, FMCG stands as the third largest industry in the country with 11% of total investments in the industrial sector (Al-Eqtisadia, 2010).

The most significant difference between the FMCG industry and the automotive industry, the "mother" of *lean* manufacturing, is the long setup and change-over time (Found & Rich, 2007). This explains the inclination to produce FMCGs in large batches. In addition to high volume production, product variability is high in FMCG manufacturing since a large portion of products falls under the 'me-too' category, i.e. the need to cover a large product portfolio just to keep up with the competition (Francis, Dorrington & Hines, 2006).

FMCG falls under the business-to-consumer (B2C) category. In this category, it is important to make a distinction between customers and consumers. Customers are typically retailers through whom goods are sold to consumers (Webster, Beach, & Fouweather, 2006). Therefore, the end customer, through whose perception Womack & Jones (2003, p. 16) define value, is in the case of FMCG the consumer (end user) and not the customer (the retailer). The relationship between the manufacturer and its customer, the retailer, is typically characterised by fierce price competition. In addition to this, the small number of retailers and the relatively large number of manufacturers give more power to buyers (retailers) over sellers (manufacturers). This pushes the manufacturers to focus on cost reduction (Webster, Beach & Fouweather, 2006), and it is for this reason that "many FMCG companies emphasise the short term gains of getting [their] stock costs down" (Steele & Plunkett, 1994, p. 16).

2.5 How lean implementation can benefit the FMCG industry

With an ongoing increase in prices of raw and packaging materials, and FMCG companies being under competitive market pressure to maintain their products' selling prices, profitability is directly threatened unless companies implement initiatives to reduce cost (Shabaan & Awni, 2014). Cost reduction through *lean* can be achieved through increased quality (i.e. fewer defects), reduced inventory, improved customer service, shortened order cycle time, improved manufacturing and supply chain visibility, improved manufacturing and supply chain flexibility, improved operational performance, increased operational capacity, shortened product development time, and workplace safety and cleanliness (Buxton & Jutras, 2006; Michaels, 1999; Pettersen, 2009; Womack & Jones, 1996).

3. METHODOLOGY

We adopted an exploratory, in-depth, qualitative multiple case-study approach for this research (White, 2000, pp. 39-42). We used a qualitative research because the study was non-numerical and descriptive, and the researchers were often part of the research (White, 2000, p. 28). The real-life experiences of the research participants in the FMCG industry (through personal interviews) provided very rich data which could not have been obtained by using a quantitative research approach. We used the case study approach because of the nature of the research question and objectives (i.e. the use of "how?" and "what?" questions) and the study's focus on contemporary, rather than historical events (Yin, 2009, pp. 8-14). We also used multiple case studies (as opposed to a single case study) to increase the validity of the results and reduce the chance of having unique conditions which may surround a single case (Yin, 2009, p. 61).

3.1 Sample Selection and Criteria

The FMCG industry is categorised into three major segments: food, beverage, and household (Key Note, 2006 cited in Francis, Dorrington & Hines, 2008). Therefore, in order to enhance the external validity of our research, we aimed to cover these three segments by building multiple case studies on three FMCG firms in Saudi Arabia operating in three different FMCG industries: beverages, snacks and packaged baked foods. Of these three cases, the Beverage Company was the main case (with 15 participants interviewed) while the second and third case studies (with three interviews each) were used to replicate the results from the first case study (Yin, 2009, p. 54). In addition, we carried out two confirmatory interviews with one interview with a participant from a Personal Care Products Company based in the UK, and another interview with a US Confectionary Company. These two interviews were aimed at expanding the research coverage. The interview with the participant from the Confectionary Company was the result of a snowballing sampling approach (Müller and Lecoeuvre, 2014), and involved interviewing one of the Beverage Company's participants based on his past experience in a Confectionary Company in the US. These two interviews were used to examine issues emerging from the three Saudi cases which appeared to be regionally based (i.e. issues that were influenced by the business environment in Saudi Arabia), with the aim of identifying if these issues were relevant to the FMCG industries in the UK and the US. This also served to improve the understanding of the findings obtained from the 3 main cases. Table 3 summarizes the number of interviews for each case study.

Table 3 - Summary of interviews

Case Study	Number of Interviews	Number of Participants	Type of Interview
Case Study 1 (Beverage)	9	9	Semi-structured
	1	6	Focus Group
Case Study 2 (Snacks)	3	3	Semi-structured
Case Study 3 (Baked Goods)	2	3	Semi-structured
Confirmatory Interview 1 (Personal Care)	1	1	Semi-structured
Confirmatory Interview 2) Confectionary)	1	1	Semi-structured

3.2 Data Collection

The total number of interviewees was 22 persons and they were interviewed in 17 different sessions comprising 16 individual interviews and a single focus

group session for 6 members of the shop floor staff of the Beverage Operations. We selected the interview participants from different functions (manufacturing, warehousing and sales), various levels (directors to shop floor staff) and different years of service (from over 12 years of service to less than a year) in the organisations under study. Different perspectives were, thus, brought to bear on the research due to the different participants' backgrounds. In presenting the results for the shop floor staff focus group session, we considered the group's feedback as a single feedback on the basis of unanimity.

Both primary and secondary data were used. Primary data collection was conducted via semi-structured face-to-face interviews in the manufacturing operations as well as observations made by the researchers during site visits. However, the interviews with a participant each from the UK-based FMCG case study and the US-based FMCG case study were made over the phone. Also a site visit to the UK and US facilities was not possible. The interviews were recorded via a digital (MP3) recorder, and later transcribed for subsequent data analysis. The average time for each interview was about 60 minutes and the focus group session was 90 minutes. The type of observation method used was the non-participant (structured) observation (White, 2000, p. 35). Observations included observing general site status in terms of inventory levels, presence of systems and overall conditions of the site. These observations were generally not influenced by the presence of the observer (i.e. the researchers) on the site. Secondary data comprised records of key performance indicators (KPIs) obtained from participating companies. The KPIs included production line efficiencies, inventory levels and forecast accuracy.

3.2 Data Analysis

The analysis of the research data (i.e. both primary and secondary data) followed the simultaneous flows of activity proposed by Miles & Huberman (1994, pp.10-11) which encompasses data reduction, data display and conclusion drawing/verification. The data was reduced by allocating chunks of data to previously set categories (Ghauri et al.,

1995). The 'set categories' were identified from the literature review prior to conducting the interviews. The display of the data was done using a matrix format via case ordered Meta-Matrix; i.e. Excel spreadsheet (Miles & Huberman, 1994, pp. 187-200). The matrix display was used to facilitate the analysis of the data by identifying themes and categories relevant to the research issues. In addition, the display was used to facilitate comparisons (including similarities and differences) between the cases to aid in drawing up conclusions.

4. RESULTS

4.1 The first research objective

The first research objective was to determine whether lean principles could be implemented in the FCMG industry. Our logic here was that if the seven types of wastes could be demonstrated as evident in the FCMG industry, then lean tools could be used to eliminate the waste. We asked the participants a question about whether lean could be applied to the industry three times during the interviews, i.e. once at the beginning (for participants with some previous knowledge of lean principles), then after completing the segment of questions related to the first research objective and finally after completing the segment of questions related to the third research objective. The rationale for this was to find out whether the participants had developed different opinions based on the discussions. Figure 1 summarizes the results. As can be seen from the figure, by the third time of asking this question, there was no participant who felt lean could not be applied to the FMCG Industry, as 94% of the participants responded 'yes' or 'to some extent', an increase of 6% over the first time the question was asked. The participant who had responded 'no' explained that it was not possible to implement lean under the current operational conditions. These conditions are discussed in more detail later on in this article.

Figure 1 - Participants' responses on the first question on the applicability of lean at different points during the interviews

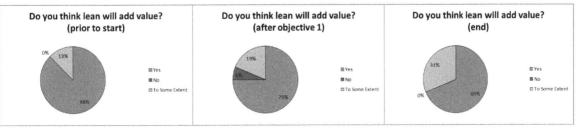

The next series of questions aimed to find out which of the seven wastes were considered by participants as present in the industry. If it could be established that wastes were present, then this would justify the need to implement *lean*. Figure 2 summarises the participants' feedback to these questions.

Figure 2 - Participants' responses on the 7 wastes

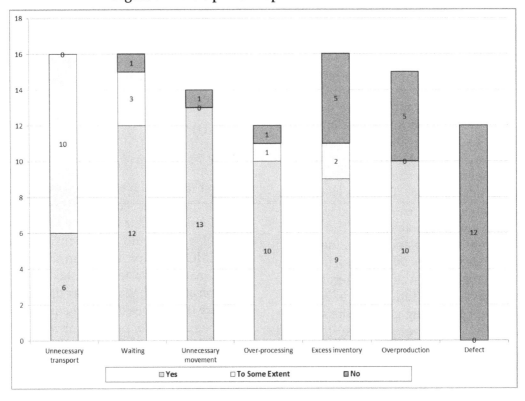

Unnecessary Transport was observed by all participants as either present 'to some extent' (63%) or as a 'common practice' (38%). In addition, we observed *Unnecessary transport* to be inherent in the design of the production lines themselves, as there were excessive stretches of conveyors built between machines. These conveyors were not used to simply transport goods, but they also acted as dynamic storage areas for the semi-processed products to protect the lines against stoppage. When we asked participants whether their shop floor staff were idle at any point in time during operation in connection with the waste of *Waiting*, 94% replied that either this was a common observation or it was observed 'to some extent' in their operations. Overstaffing was used to cover absenteeism and to accommodate the peak hours of the day in all three operations. 81% of participants felt that *Unnecessary movement* of shop floor staff was definitely present in their operations, while eleven participants (69%) believed that *Over-processing* was either 'definitely' or 'to some extent' present in their operations due to gaps in the process design and also facility layout restrictions. Most of the participants (68%) felt that the average in-

ventory carried in their operations was excessive and could be reduced, suggesting therefore, that *Excess Inventory*, is a common type of waste. In the Beverage Company and Snacks Operations, this was related to both raw materials and finished goods inventory, while in the Baked Foods Operations, this was only related to raw materials since the average inventory of finished goods ranged from 12 to 24 hours only. While *Overproduction* received 77% positive responses from the participants from the Beverage and the Snacks Companies, it was not considered to be a problem in the Packaged Baked Foods Company. This appeared reasonable since the Baked Foods manufacturer produced only according to customer orders, which was largely due to the relatively short shelf life of its products in comparison with the Beverage and Snacks manufacturers (i.e. a maximum of 45 days in Baked Foods as against 4 months for Snacks and up to 12 months for Beverages). *Defect* was the only type of waste which was systematically monitored in all participating companies' operations in the form of defective raw materials and finished goods before, during and after production. None of the participants was of the view that the company's percentage of defects was high. This was also confirmed by the participating companies' internal quality records showing benchmark results both in terms of trade product quality results and in-house defect rates.

During the interviews, we asked the participants about what they believed were the causes of these types of wastes in their operations. Figure 3 summarizes their feedback, which has been categorized according to the type of waste.

Figure 3 - Causes of the seven types of wastes

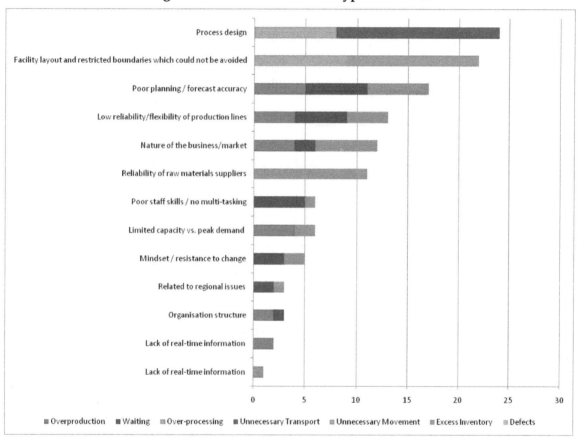

Deficiencies in process design and facilities layouts were the causes most frequently quoted by participants for the wastes. These, however, were mainly associated with two types of wastes: *Unnecessary Transport* and *Unnecessary Movement*, i.e. the first and third rated

types of waste. A close look at the causes, however, shows that reliability of raw material suppliers was the cause cited by the largest number of participants. Overall, the top five causes of waste as mentioned by the participants are as shown in Table 4.

Table 4 - Top 5 causes of waste as cited by participants

Top Five Causes of Waste	Number of Participants	Number of Operations
Reliability of raw materials suppliers	11	3
Low reliability/flexibility of production lines	9	3
Poor planning/forecast accuracy	9	3
Nature of business/market dynamics	8	3
Facility layout and restricted boundaries which could be avoided	7	3

4.2 *The second research objective*

The second research objective aimed at examining the conditions under which *lean* principles could be applied in the FMCG industry's operations. In order to understand the conditions which needed to be changed, we asked the participants about which conditions they felt needed to be changed prior to implementing *lean*. Responses to this question varied from one operation to another. While forecast accuracy and supplier reliability appeared to be major issues in the Beverage Operation, mind-set appeared to be the issue in the Snacks operations. Table 5 summarizes the prerequisites to *lean* implementation in FMCG. Its aim is to provide an overview of important aspects and not to make inferences about the issues raised.

Table 5 - Prerequisites to lean implementation in FMCG

Prerequisites to *Lean* Implementation	Number of participants citing this prerequisite	Equivalent % from total participants
Improve planning / forecast accuracy	6	38%
Invest in automation & IT systems	6	38%
Get top management support	4	25%
Improve supplier reliability	3	19%
Improve production line reliability	3	19%
Increase skill level of floor shop staff	3	19%
Improve facility design/layout	3	19%
Change of mind-set	1	1%

4.3 *The third research objective*

The third research objective aimed to identify which *lean* tools and techniques could be applied to aid a *lean* implementation in the FMCG industry.

4.3.1 *Just-In-Time (JIT) tools:*

Kanban, the basis of just-in-time, was fairly acceptable with 73% of participants indicating that this tool could be applied either 'fully' or "in some areas but not others' in their operations. However, further analysis shows that the percentage is skewed upwards mainly by the 3 participants from the Baked Foods manufacturer, who all agreed that *Kanban* can be fully implemented in their operations. This was the case because the downstream operations of the company were already just-in-time (i.e. based on actual customer orders). Participants from the two other operations (Beverage & Snacks) were equally split between 'Yes' (33%), 'No' (33%) and 'To Some Extent' (33%).

Andon was also a point of debate with 53% rejecting it, 40% finding it a useful tool and one participant (7%) saying that it was a good tool but should be implemented with care to avoid excessive stoppages on the production line, as this could in turn result in serious quality issues and increased losses.

4.3.2 *Autonomation tools:*

The automatic stoppage of the production line (i.e. *jidoka*), received more acceptance (73%) than the manual stoppage (i.e. *andon*). This was mainly driven by 100% acceptance rate from the Beverage Company. The exception was the Beverage company's Warehouse Manager who indicated that it was not applicable in his operations since there was no automation in his area yet.

4.3.3 *Production Levelling Tools:*

Takt-time was generally dismissed by 79% of the participants. Only the Warehouse Managers of both the Beverage Company and the Snacks Operations indicated that it was applicable in their manual warehousing operations, but not in automated manufacturing operations. Similarly, the Plant Manager of the Baked Foods Operations, saw it as being only applicable in the heavily manual packing area in his plant. Most participants, on the other hand, did not see any value being derived from this tool.

Heijunka, or production levelling, was discussed from the perspective of the ability to level demand, since in order for a business to be able to level its production, it must be able to level its market demand (Ohno, 1988, pp. 36-38), i.e. level demand across the various seasons of the year and across the various weeks of a month. 44% of the participants did not think this was possible since market demand was seasonal, and this was especially the case in the Baked Foods operations where products must be produced and delivered to customers within their very short life (as short as 7 days in some cases). This means that building up inventory in low-demand months to cover the demand during high-demand months would not apply (and this would also defy the purpose of 'just-in-time' production).

In discussing *Quick Changeover*, participants were asked whether changeover time – which averaged from three to five hours – could be reduced to 30 minutes or less. 90% of participants responded with either 'yes' or 'to some extent'.

A senior participant from the Beverage Company was of the view that applying the concept of small, frequent production runs (which would result from using the quick changeover concept) would be a challenge in the beverage industry because the process of changing over from one batch of syrup to another generates waste in syrup, concentrate and sugar, no matter what the size of the run is; meaning that losses will be heavier if run sizes were smaller.

4.3.4 Process Improvement Tools:

Both *Kaizen* & *value stream mapping* were highly acceptable tools with *Kaizen* receiving 100% acceptance rate, and *value stream mapping* 90%. By contrast, *Kaikaku* (which involves one-time major changes as opposed to continuous small changes in *Kaizen*), was not found to be a useful tool by 60% of participants. However, there were significant variations between the different operations. While the tool received 100% acceptance by the Snacks Operations' participants, it was totally dismissed by the Baked Foods participants. This was explained by one of the operation's participants as being due to "*the nature of our products and the way it is manufactured. They are fixed manufacturing lines*". Similarly 70% of the Beverage Company's participants were of the view that the tool was not helpful to their operations. This appeared to be influenced by the company's plan to

build a new manufacturing facility to replace the existing facilities, and this discouraged the will to make major investments to implement changes in the current facilities.

Visualization Tools:

5S and *poke yoke* were generally accepted tools that could be easily implemented in any area of the operation. *5S*, which is already implemented in the Beverage Company's manufacturing operations, received a 100% acceptance rate. Similarly *poke yoke* was found to be a useful tool by all participants, with the exception of the Warehouse Manager of the Beverage Company who indicated that it was not applicable to his manual warehouse operations.

In general, the following factors that differentiate between the packaged goods industry from the automotive industry (where *lean* first started) were mentioned by the participants as obstacles against the use of some *lean* tools and techniques in the beverage industry:

a. Many ingredients in beverage, snacks or baked goods are perishable and would decay or rot if production is stopped frequently. Furthermore, frequent stoppages and start-ups on the continuous production lines would result in degradation of the product quality. This becomes a challenge against the use of *andon* in the industry.

b. Some market mechanics do not allow waiting for consumer (end customer) orders prior to finalizing the production plans. These mechanics include the following:

• The consumer's decision to purchase is typically done on spot and is not planned ahead. For instance, the Beverage Company's records showed that over 60% of their sales are made to small and medium sized grocery stores, where consumers' decision to purchase are typically spontaneous. If the consumer does not find, for instance, Cola brand A, he/she will simply shift to Cola brand B. It would not be possible to wait for confirmed customer/consumer orders to finalise the production plan. Furthermore, it is not possible to produce all of the company's 120+ products every day. For this reason keeping some inventory in the plant and warehouses and using sales forecasts is considered inevitable. These issues pose a challenge to implementing *kanban* and *takt-time*.

- Product sales are seasonal. This may also be the case for cars sales as well. However, unlike car purchase, which typically happens once every few years, the purchase of FMCGs is a routine, or even a daily, habit. As highlighted by some participants, consumer promotions typically aim to divert consumers from one brand to another, but do not encourage consumers to forward-buy their requirements in the low sales season several months ahead of the time of consumption. Similarly, sales tend to increase towards month end in the Saudi market. This is because consumers there tend to do bulk buying when they receive their monthly pay checks. These two factors can make it difficult to level demand, a key prerequisite for production levelling (*heijunka*) (Ohno, 1988, pp. 36-38).

4.4. The fourth research objective

The fourth research objective aimed to identify the benefits and drawbacks of implementing *lean* in the operations under study. To achieve this, participants were asked about what they considered would be the benefits of implementing *lean* in their manufacturing and warehousing operations. Table 6 summarises the benefits as foreseen by participants.

Table 6- Foreseen benefits of lean implementation in FMCG

Benefits of *Lean* Implementation	Number of participants citing this benefit	Equivalent % from total participants
Improving performance efficiency	10	63%
Reducing waste	6	38%
Reducing cost	6	38%
Reducing manpower	6	38%
Reducing inventory & storage space	5	31%
Improving quality	4	25%
Improving customer service	4	25%
Improving staff skills	2	13%
Improving performance of raw materials suppliers	1	6%
Improving work environment	1	6%
Healthy growth of business	1	6%

Finally, on the drawbacks of implementing *lean* in FMCG operations, while most participants did not foresee any drawbacks, some participants highlighted a few 'watch-outs'. These are summarized in Table 7. Similar to Table 5, the aims of Tables 6 and 7 are to provide an overview of important aspects and not to make inferences about the issues raised.

Table 7- Drawbacks of lean implementation

Drawbacks of *Lean* Implementation	Number of participants citing this drawback	Equivalent % from total participants
If not implemented correctly, could backfire	5	31%
Some lean tools are not suitable for the industry	2	13%
Some employees would lose their jobs	1	6%
Would require some investment in training to ensure success	1	6%

4.5 Summary of Results for the Two Confirmatory Interviews

This section looks at the key findings from interviews with a participant each from the UK operations of a multinational Personal Care Products Company, and the US operations of a multinational Confectionary manufacturer. The issues are mainly those emerging from the three main case studies which were considered as regionally-based, i.e. influenced by the business environment in Saudi Arabia. Focus is therefore given to the areas which participants from the three main case studies highlighted as related to the Saudi Arabia region including *Waiting* and *Excess inventory*.

4.6 Personal Care Products – UK operations

The company is already implementing a Total Productive Maintenance (TPM) system, which is, like *lean,* based on Japanese principles, and thus, has many intersecting areas with *lean* such as *kaizen* and *5S* (Venkatesh, 2009). Products typically had a long shelf life (as long as 24 months). Raw materials were delivered just-in-time from either nearby suppliers, or even onsite suppliers. The company produced against forecasts, not actual customer orders, but nonetheless, forecast accuracy level was high by having the forecast refined by customers. The types of waste seen as common in the operations were: *Waiting* (to some extent), *Excess inventory* (in finished goods only), and *Unnecessary movement. Waiting* had been significantly reduced in the operations, but not totally eliminated. Although raw materials were delivered just-in-time, the average inventory of finished goods was two weeks. The interview participant was of the view that this level of inventory was not *Excess inventory,* since this inventory was used as a buffer to protect against fluctuations in market demand. *Unnecessary movement* was driven by lack of automation in some areas of the operation, resulting in the use of labour to move materials. This, however, had been corrected in some plants but not in all. The interview participant agreed that all *lean* tools under study would be useful to a *lean* implementation. The only exception, however, was *takt-time,* which he explained as follows: *"Because what we have is a continuous process. We have a minimum time to fill the bottle".*

4.6.1 Confectionary – US operations

This operation was not originally planned as part of the study but was based on a snowballing sampling approach, i.e. a referral by the Project Manager of the Beverage Company (Müller and Lecoeuvre, 2014). The key findings were that production was made against forecasts, not actual customer orders. However, production was made only to the exact forecast quantity without any buffers. As a result, finished goods inventory in the plant was a maximum of two days. Raw materials inventory, on the other hand, was much higher than that, particularly for packaging materials. The types of waste seen as common in the operations were *Waiting* and *Excess inventory* (in raw materials only). *Waiting* was driven by variations in demand and the participant explained this as follows: *"We used to have waiting time. Like, for example, production demand dropped. Sales could not sell it, so we used to have 4 out of 5 days, or actually 2 out of 3 weeks production".* This finding also suggested a need for the productive levelling tool, *Heijunka. Excess inventory* was the case only for raw materials, and was mainly because of the vast distances in the US, where suppliers were typically located far from the plant, which could be several states away. *Value stream mapping* and *kaizen* were cited by the participant as systems that were already in place in the Confectionary Operation.

5. DISCUSSION

5.1 The First Research Objective

Although about half of the participants were of the view that *lean* could not be implemented in their operations under their current conditions, while a third of the participants were of the view that *lean* principles could be implemented only partially in their operations, the participants were unanimous in agreeing that *lean* could be applied to their operations. The strong conviction by the participants that implementing *lean* would add value to their operations appeared to be due to the realisation that six of the seven wastes were found to exist in all of the five operations studied. These wastes, ranked from the most to the least as cited by the participants, were: *Unnecessary transport, Waiting, Unnecessary Movement, Over-processing, Excess inventory* and *Overproduction.* Although *Defect* was also present, it was already being monitored by all the operations, and therefore was not considered a justification for *lean* implementation.

The causes of the wastes as suggested by the participants were found to be in line with those identified in the *lean* literature review (for example, Liker, 2004, pp. 28-29). However, these suggested that the causes could not be generalised to the FMCG industry as a whole. For instance, the results suggested that supplier reliability, which was seen as the main cause for *Excess inventory*, would appear to be a regional issue since this was not seen as an issue by the participants from the UK-based case study and the US-based case study. Similarly, poor planning/forecast accuracy, which was associated with *Overproduction, Waiting* and *Excess inventory*, was seen as an issue by the participants from the Beverage, Snacks and Confectionary operations, and appeared to be a result of the produce-to-forecast model adopted by these operations. However, this was not an issue in the Baked Foods Operations, where a produce-to-order model was adopted. It was also not seen as an issue in the UK-based personal care products operation although this operation also adopted a produce-to-forecast model, mainly because the company operated an advanced forecasting system where forecasts were developed in line with the customers.

Furthermore, a review of the literature reveals that supplier reliability (which was considered as one of the key causes of waste in this research) is not limited to the Saudi Arabian market. For instance, a study by Hindricks & Singhal (2003), on several multinational firms including Sony, Nike and Ericsson concluded that supply chain disruptions resulting from supplier failures had resulted in an average of 10% decrease in shareholder returns. Similarly, the literature on the FMCG industry such as Adebanjo & Mann (2000), suggests that demand consumer forecasting (which was considered as another key cause of waste by the research participants) is an integral part of the industry and that effective forecasting would result in increased product availability to consumers and lower inventory. Therefore, this study suggests that both supplier reliability and poor planning issues and their subsequent wastes are wastes that generally exist in the FMCG industry.

The results also suggest that the issue of facility layout and restricted boundaries were specific to the operations under study and could not be generalised to the industry as a whole. This was highlighted several times during the interview with the Baked Foods Operation's Plant Manager, who after pointing out the restrictions in his operation clarified that "*this is not the way the industry should be*". It could still be argued, nevertheless, that these types

of facility restrictions were typically a result of capacity expansions within existing facilities, which is a very common practice in all industries. Kulturel-Kunak (2007), for example, estimates that more than $250 billion are spent in the US annually on facility layout redesign. This is also supported by the finding that the majority of 4 out of 5 case studies in this research highlighted facility layout restrictions as an issue in their operations.

The view by a third of the participants that *lean* could only be partially implemented in their operations appeared to be mainly driven by one of the major contributors to waste in the FMCG industry; i.e. the 'nature of the business/market'. Under this broad category falls several differences in product characteristics and market dynamics which differentiate the FMCG industry from the automotive industry such as the perishable nature of the products and their ingredients, the spontaneous purchase by consumers and the seasonality of sales trends. This would therefore suggest a need for a customised approach to *lean implementation*, which is also in line with the findings of Bicheno & Holweg (2009, pp. 50-51), who proposed that, depending on the type of industry, a product-process matrix could be used to determine the best way to implement *lean* in an industry. For instance, the matrix recommends that for a high-volume/continuous manufacturing operation such as the FMCG, a *lean* implementation could be best supported by optimisation scheduling tools such linear & mathematical programming (Bicheno & Holweg, 2009, pp. 50-51).

In summary, in relation to the first research objective, all participants agreed that *lean* could be applied in their FMCG operations, which was not surprising since the same types of waste identified by Ohno (1988, pp. 19-20) for the auto industry were seen to be present in the participating FMCG operations as well.

5.2 The Second Research Objective

Responses from the participants suggested that prior to starting a *lean* implementation in their operations some conditions needed to be changed. These are categorized in Table 8.

Table 8 - Prerequisites to lean implementation in FMCG

Category	Conditions Under This Category
Management	• Securing top management support
Human Resources	• Setting up a structured job rotation scheme in place • Providing *lean* training to staff to overcome *resistance to change.*
Technology	• Investing in information technology • Investing in plant automation
Process Improvements	• Improving sales planning process

The statement that management support is a prerequisite to a successful *lean* implementation is very much in line with the findings of Moyano-Fuentes & Sacristan-Diaz (2012) who carried out a thorough review of *lean* literature and highlighted that although managerial commitment was not considered until fairly recently, it is of key importance to the success of any *lean* implementation. The authors made references to several studies in a number of different industries which identified lack of management support as the first obstacle when implementing *lean*. The prerequisites under the Human Resources (HR) category are in line with the statement by Bhasin (2012, p. 439) that *lean* literature "dictates that nine of the top ten barriers to change are quoted as being people-related, including poor communications and employee opposition." This is also in line with Shah & Ward (2003 cited in Furlan, Venilli & Dal Pont, 2011) who segmented *lean* implementation into three *lean* bundles, JIT, TQM and HRM, and argued that these three bundles are interdependent and must each be successfully implemented to ensure the success of any *lean* implementation. HRM is linked to the HR aspects on *lean* implementation.

However, while the conditions related to 'Technology' and 'Process Improvement' in Table 7 are suggested as pre-requisites for *lean* implementation, it could however, be argued that these conditions would actually occur as a *result of* rather than as a *prerequisite to lean* implementation. This statement is supported by the *lean* implementation action plan by Womack & Jones (2003, pp. 247-271). Thus, actually, these conditions are not required to be in place prior to implementing *lean*, as they would rather come into existence as a result of implementing *lean*.

Studies on *lean* have identified similar factors as influencing its implementation, several of which intersect with the findings of this research. For example, in a study to develop a full-blown conceptual model for *lean* implementation, Roslin & Shahadat (2014)

identified the following to be 'influential factors' in implementing *lean* manufacturing: management commitment, employee empowerment, employee involvement, HR management, teamwork, organizational change, customer relationship management, supplier relationship management, information technology.

5.3 The Third Research Objective

The results suggested that the visualisation tools (i.e. *poke yoke* and *5S*) and the process improvement tools (i.e. *value stream mapping, kaizen* and *kaikaku*) were generally more accepted than the production levelling and JIT tools, which is one of the two pillars of the Toyota Production System (Ohno, 1988, p. 4). This high acceptance rate for the visualisation and the process improvement tools appears to agree with the findings of Abdullah (2003) that *5S* (one of the visualisation tools) and *Value Stream Mapping* (one of the process improvement tools) are "universally applicable", whereas the degree of applicability of other tools depends on the specific characteristics of the process industry under study.

The autonomation tool, *jidoka*, received a high acceptance rate as well with 87% of participants saying that it could be a useful tool in their production operations since their manufacturing consisted of automated, high-speed production lines, which suggests that an automated system would be more capable of capturing errors and defects than humans.

The dilemma, however, was that although participants provided good justifications for rejecting some of the JIT and production levelling tools, the above statements would seem to be paradoxical. This is because, on the one hand, participants appeared to be saying that *lean* could be implemented in their operations, while on the other hand they appeared to be negating this same statement by saying that basic *lean* tools such as *andon, takt-time* and *heijunka* were not suitable for their operations. However, a critical analysis of the research results shows that several of the seven types of wastes found in the industries studied were attributed to the 'nature of business/ market'. While a customer buying a car would be willing to wait for several days and even weeks to get the car required, a FMCG consumer would certainly not be willing to wait, even for a few minutes, to get their specific cold refreshing drink, pack of potato chips or bag of bread from the retail outlet, but would simply buy the next available brand on the shelf. Therefore, while Toyota could afford to wait for their customer orders prior to making their final production plans on the assembly line (Ohno, 1998, pp. 48-50), most FMCG manufacturers would have no choice but to rely heavily on sales forecasts based on historical trends. For instance, the Beverage Company's records showed that, historically, forecast accuracy was in the region of 80% for total sales, and 50% by product. These results, although considered good accuracy results as per the Multinational Principal's guidelines, are far from being as accurate as a 'produce-to-order' model. It would appear the participants who rejected kanban might have had this preconception in mind, which probably explains why 'improving planning/forecast accuracy' was on top of the list of conditions to be changed prior to lean implementation as was suggested by the Beverage Company's participants. However, this preconception would appear to be false since the Baked Foods Operation was successful in planning their production based on actual sales orders from customers. Thus, this would suggest the need for a change of mindset in order to change the way business is done in FMCG industry manufacturing and warehousing operations, as also supported by the findings of the second research objective. A post-interview discussion with the Product Availability Manager from the Beverage Company confirmed this point. He stated that many FMCG companies have actually switched to make-to-order model instead of make-to-forecast.

Therefore, it could be concluded that with the exception of andon and heijunka (for the reasons previously explained), all of the lean tools which were discussed could be implemented in FMCG industries and would support the success of a broader lean implementation across the business.

5.4 The Fourth Research Objective

The research participants identified several benefits of implementing lean in their operations, including

the following: improved performance with respect to efficiency, quality and customer service, reduced waste with respect to cost, manpower and inventory, and improved quality of staff and suppliers. These were in line with those suggested in the lean literature (for example, Buxton & Jutras, 2006; Michaels, 1999; Pettersen, 2009; Womack & Jones, 1996) and it goes to stress the point that reducing cost is actually what lean is all about. As noted by the creator of the Toyota Production System, Ohno (1988, p. 9), "Cost reduction must be the goal of consumer products manufacturers trying to survive in today's market place." Another expected benefit from lean implementation was inventory reduction or elimination. However, although it was strongly accepted that a significant reduction of inventories (both raw materials and finished goods) could be achieved, none of the participants stated that eliminating inventory altogether was possible in the FMCG industry. This seemed reasonable considering the relatively high level of demand uncertainty in the FMCG industry. Womack, Jones & Roos (2007, p. 295) made a similar observation with respect to Toyota in the afterward of their third edition of The Machine That Changed the World that:

> In fact, Toyota's JIT is far from a zero inventory system because some inventory, what Toyota calls "standard inventory" is still needed in practically every production process, typically near the end. This inventory is proportional to the volatility of orders from the customer and to the stability of the upstream process steps delivering what is needed.

On drawbacks of implementing lean, there were no significant issues. However, some participants cautioned that if the implementation of lean was not properly carried out, it could result in a decline in performance and an increase in waste, a point also highlighted by several scholars like Bicheno & Holweg (2009, pp. 44-48), Liker (2004, pp. 10-14), Ohno (1988, pp. 29-30) and Womack & Jones (1996).

5.5 A Conceptual framework for the Application of Lean in the FMCG Industry

From the foregoing discussion, a conceptual framework for the application of lean by the FMCG Industry is proposed in Figure 9. The lean tools found to be appropriate for the FMCG operations (in Research Objective 3) are used to assist in eliminating the wastes found to be present in the operation, i.e. Overproduc-

tion, Waiting, Unnecessary transport, Over-processing, Excess Inventory, and Unnecessary Movement (in Research Objective 1). This should be done under the conditions identified as essential to be in place prior to a success-ful *lean* implementation (i.e. in Research Objective 2) to ensure a successful *lean* implementation. The out-comes of the successful *lean* application are the benefits and drawbacks (in Research Objective 4).

Figure 4 - A Conceptual Framework for Lean Implementation in The FMCG Industry

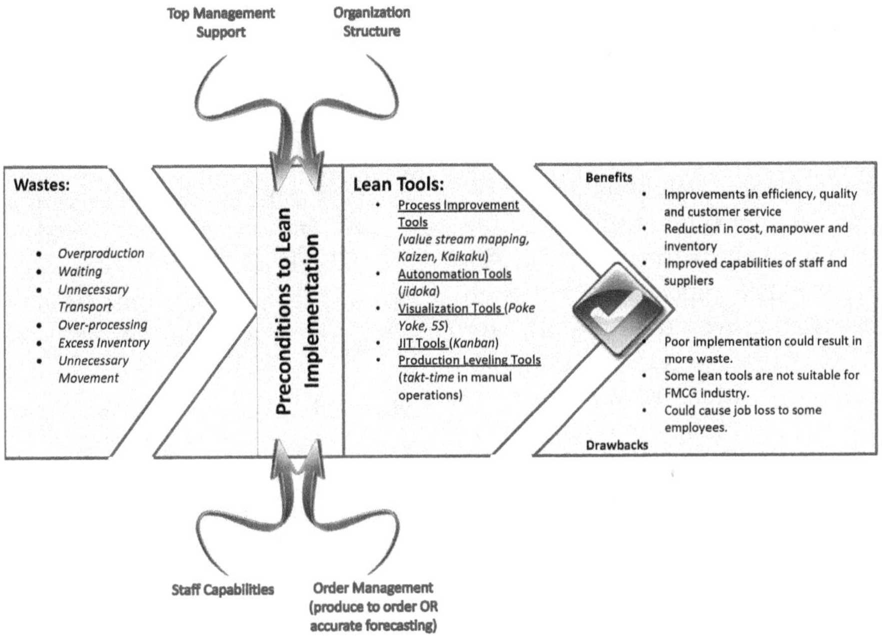

6. CONCLUSION

This research aimed to cover a gap in the *lean* literature with respect to the applicability of *lean* principles in the FMCG industry from a study of five different FMCG operations. These are: a Beverage Operation, a Snacks Operation, a Packaged Baked Goods Operation, a Personal Care Products Operation and a Confectionary Operation.

Six of the seven types of waste, which were identified in the automotive industry where the *lean* system originated, were found to be also present in the FMCG operations studied. The seventh waste, *Defect*, was also present but was already being monitored by all the operations, and therefore was not considered a justification for *lean* implementation. The identification of these wastes justifies the conclusion that *lean* implementation in FMCG Industry is possible and justifiable to eliminate these wastes and add value. However, due to differences in the 'nature of business' between the automotive and FMCG industries a customised approach including the use of optimisation scheduling tools like linear & mathematical programming (Bicheno & Holweg,

2009, pp. 50-51) would be required to implementing *lean* in the FMCG industry.

Several conditions were found to need changing prior to starting a *lean* implementation in the FMCG operations. These were categorized into four groups: Support, Human Resources, Technology and Process Improvements. Support and Human Resources are genuine prerequisites and were found to be in line with the *lean* literature. It was argued that Technology and Process Improvements, on the other hand, are actually not prerequisites to *lean* implementation but rather steps resulting from the implementation process.

While many of the visualisation and process improvement tools (*5S, poke yoke, value stream* mapping and *kaizen*) were, by their general nature, found to be applicable in the FMCG industry, some of the more specialised just-in-time and production levelling tools could actually end up generating more waste. It is, therefore, important that *lean* tools should not be applied blindly in the FMCG industry but the most appropriate tools should be selected based upon net benefits for the FMCG industry.

Several benefits expected to be derived from implementing *lean* in the FMCG industry manufacturing and warehousing were identified. The most important of these were improvements in performance, quality and customer service and reductions in cost, manpower and inventory. No significant drawbacks were expected from *lean* implementation. Nevertheless, it is possible there could be more waste generation, and job loss for some employees if *lean* is not properly implemented.

This study contributes to filling a gap in the *lean* literature on the applicability of lean principles in the FMCG manufacturing and warehousing operations. The study shows that the argument for the universality of *lean* for all industries (Ohno ,1988, p. 9) appears valid for FMCG operations, although some level of customisation would be required for *lean* to benefit FMCG manufacturers. In addition, the research identifies conditions required for the successful implementation of *lean*, the implementation approach, and also the benefits and drawbacks. Furthermore, we propose a suitable conceptual framework that could be used as a guide for the implementation of *lean* in the operations of FMCG manufacturing organizations.

Although we adopted a qualitative research approach which has the limitation of generalizing the findings, we made the effort to minimize this limitation by employing a multiple case study involving three main categories of the FMCG industry, i.e. food, beverages and household products. In addition, the inclusion of two other cases, one from the UK (Personal Care Products) and the other from the US (Confectionery Products) sought to broaden the otherwise 'regional outlook' of the study and enhance its robustness and external validity.

While the results of this research suggest that *lean* could be implemented in the FMCG operations, further research is still required. In particular, an area for future research could be to test the statistical generalisability of the conclusions from this study. Such research could adopt quantitative methods of data collection by utilising the variables (i.e. the themes) generated in this research and the findings as hypothesis, to provide more external validity to the outcome of this research, and probably also help to refine the conceptual framework so that its usefulness as a tool for theory, practice and policy making may be maximised.

REFERENCES

Abdullah, F. (2003) *Lean Manufacturing Tools and Techniques in the Process Industry with a Focus on Steel*, University of Pittsburgh [Online]. Available from: http://etd.library.pitt.edu/ETD/available/etd-05282003-114851/unrestricted/Abdullah.pdf (Accessed: 17 December, 2010).

Adebanjo, D. Mann, R. (2000) 'Identifying Problems in Forecasting Consumer Demand in The Fast Moving Consumer Goods Sector', *Benchmarking: An International Journal*, 7(30), pp. 223-230.

Al-Eqtisadiah (2010), 25 December [Online]. Available from: http://www.aleqt.com/2010/12/25/article_482821.html (Accessed: 03 January, 2011).

Bhsain, S. (2012) 'An appropriate change strategy for lean success', *International Journal of Operations & Production Management*, 50(3), pp. 439-458.

Bhamu, J. & Sngwan, K. S. (2014) 'Lean manufacturing: literature review and research issues', *International Journal of Operations & Production Management*, 34(7), pp. 876-940.

Bicheno, J. & Holweg, M. (2009) *The Lean Toolbox*, 4th edition. Buckingham: PICISIE Books.

Bunkley, N. (2009) 'Toyota Ahead of GM Sales in 2008 Sales', *The New York Times*, 21 January [Online]. Available from: http://www.nytimes.com/2009/01/22/business/22auto.html (Accessed: 21 July, 2010).

Buxton, m. & Jutras, c. (2006) *The Lean Supply Chain Report*, Aberdeen Group [Online]. Available from: http://www.sap.com/industries/automotive/pdf/BWP_Aberdeen_Supply_Chain_Report.pdf (Accessed: 11 June, 2010).

Chen, J, Li, Y & Shady, B. D. (2010) 'From value stream mapping toward a lean/sigma continuous improvement process: an industrial case study', *International Journal of Production Research*, 48(4), pp. 1069-1086.

Childerhouse, P. & Towill, D. R. (2004) 'Reducing Uncertainty in European Supply Chains', *Journal of Manufacturing Technical Management*, 15(7), pp. 585-598.

Cooney, R. (2002) 'Is "Lean" A Universal Production System? Batch Production in The Automotive Industry', *International Journal of Operations & Production Management*, 22(10), pp. 1130-1147.

Duguay, C. R., Landry, S. & Pasin, F. (1997) 'From Mass Production to Flexible/Agile Production', *International Journal of Operations & Production Management*, 17(12), pp. 1183-1195.

Found, P. & Rich, N. (2007) 'The Meaning of Lean: Cross Case Perception of Packaging Businesses in The UK's Fast Moving Consumer Goods Sector', *International Journal of Logistics: Research & Application*, 10(3), pp. 157-171.

Francis, M. (2004) *Lean Information and New Product Introduction: An FMCG Case Study*, BNet [Online]. Available from: http://jobfunctions.bnet.com/abstract.aspx?docid=312599 (Accessed: 6 June, 2010).

Francis, M., Dorrington, P. & Hines, P. (2006) 'Supplier Led NPD Process Improvement in the UK FMCG Industry: Need and

Method', *International Journal of Innovation*, 12(2), pp. 195-222.

Furlan, A., Vinelli, A. & Dal Pont, G. (2011) 'Complementarity and lean manufacturing bundles: an empirical analysis', *International Journal of Operations & Production Management*, 31(8), pp. 835-850.

Ghauri, P., Gronhaug, K. and Kristianslund, I., (1995), *Research Methods in Business Studies: A Practical Guide*, Prentice Hall, Hemel Hempstead.

Grant, D. B. & Banomyong, R. (2010) 'Design of closed-loop supply chain and product recovery management for fast-moving consumer goods', *Asia Pacific Journal of Marketing*, 22(2), pp. 232-246.

Hines, P. & Rich, N. (1997) 'The Seven Value Stream Mapping Tools', *International Journal of Operations & Production Management*, 17(1), pp. 46-64.

Hines, P., Holweg, M. & Rich, N. (2004) 'Learning to Evolve', *International Journal of Operations & Production Management*, 24(10), pp. 994-1011.

Howison, J. D. (2009) 'A Tough 'Cell': Implementing Lean Production at Toledo Jeep', *Critical Sociology*, 35(5), pp. 689-696.

James-Moore, S. M. & Gibbons, A. (1997) 'Is Lean Manufacture Universally Relevant? An Investigative Methodology', *International Journal of Operations & Production Management*, 17(8), pp. 899-911.

Jones, D. T., Hines, P. & Rich, N. (1997) 'Lean logistics', *International Journal of Physical Distribution & Logistics*, 27(3/4), pp. 153-173.

Kulturel-Kunak, S. (2007) 'Approached to Uncertainty in Facility Layout Problems: Perspectives at the Beginning of the 21st Century', *Journal of Intelligent Manufacturing*, 18(2), pp. 273-284.

Lean Enterprise Institute (2009) *The Machine That Changed The World – Revised* [Online]. Available from: http://www.lean.org/Bookstore/ProductDetails.cfm?SelectedProductID=160 (Accessed: 21 July, 2010).

Liker, J. K. (2004) *The Toyota Way:14 Management Principles From The World's Greatest Manufacturer*. New York: McGraw-Hill.

Michaels, L. M. J. (1999) 'The Making of a Lean Aerospace Supply Chain', *Supply Chain Management: An International Journal*. 4(3), pp. 135-145.

Miles, M. B. & Huberman, M. (1994) *Qualitative Data Analysis: An Expanded Sourcebook*. 2nd Ed, Thousand Oaks: SAGE, Inc.

Moyano-Fuentes, J. & Sacristan-Diaz, M. (2012) 'Learning on lean: a review of thinking and research', *International Journal of Operations & Production Management*, 32(5), pp. 551-582.

Müller, R. and Lecoeuvre, L (2014), 'Operationalizing governance categories of projects', *International Journal of Project Management*, Volume 32, Issue 8, pp. 1346-1357.

Mustafa, S., Dumrak, J. & Soltan, H. (2013) 'A framework for lean manufacturing implementation', *Production & Manufacturing Research: An Open Access Journal*, 1(1), pp. 44-64.

Ohno, T. (1988) *Toyota production System: Beyond Large-Scale production*. Boca Raton: Taylor & Francis Group.

Papahristodoulou, C. (1994) 'Is Lean Production The Solution?', *Economic and Industrial Democracy*, 15(3), pp. 457-476.

Pettersen, J. (2009) 'Defining Lean Production: Some Conceptual and Practical Issues', *The TQM Journal*, 21(2), pp. 127-142.

Ravikumar, M. M., Marimuthu K. & Chandramohan, D. (2009) 'Implementation of lean manufacturing in automotive manufacturing plant [TEL]', *International Journal of Applied Engineering Research*, 4(10), pp. 2041-2050.

Roslin, E. N. & Shahadat, S. A. M. (2014) 'A Conceptual Model for Full-Blown Implementation of Lean Manufacturing System in Malaysian Automotive Industry', *Proceedings of the 2014 International Conference on Industrial Engineering and Operations Management Bali, Indonesia, pp. 1309-1315.*

Shabaan, M. S. & Awni, A. H. (2014) 'Critical success factors for total productive manufacturing (TPM) deployment at Egyptian FMCG companies', *Journal of Manufacturing Technology Management*, 25(3), pp. 393-414.

Simons, D. & Zokaei, K. (2005) 'Application of lean paradigm in red meat processing', *British Food Journal*, 107(4), pp. 192-211.

Spear, S. & Bowen, H. K. (1999) 'Decoding the DNA of the Toyota Production System', *Harvard Business Review*, 77(5), pp. 96-106.

Steele, W. & Plunkett, K. (1994) 'Finished Stock: The Piggy in the Middle', *Logistics Information Management*, 7(6), pp. 16-22.

Towill, D. R. (2007) 'Exploiting the DNA of the Toyota Production System', *International Journal of Production Research*, 45(16), pp. 3619-3637.

Venkatesh, J. (2009) *An Introduction to Total Productive Maintenance (TPM)*, Plant Maintenance Resource Center [Online]. Available from: http://www.plant-maintenance.com/articles/tpm_intro.shtml (Accessed: 15 December, 2010).

Webster, M., Beach, R., & Fouweather, I. (2006) 'E-business strategy development: FMCG sector case study', *Supply Chain Management*, 11(4), pp. 353–362.

Womack, J., Jones, D. & Roos, D. (1990) *The Machine That Changed The World*. New York: Free Press.

Womack J., Jones, D. & Roos, D. (2007) 'Afterword 2007' in *The Machine That Changed The World*, 2007 edition. New York: Free Press.

Womack, J. & Jones, D. (1996) 'Beyond Toyota: How to Root Out Waste and Pursue Perfection', *Harvard Business Review*, 74(5), pp. 140-158.

Womack, J. & Jones, D. (2003) *Lean Thinking*. New York: Free Press.

Wood, N. (2004) *Lean Thinking: What is it and What It Isn't*, Management Services [Online]. Available from: http://www.allbusiness.com/management/960002-1.html (Accessed: 7 July, 2010).

Yin, R. K. (2009) *Case Study Research: Design and Methods*. 4th Edition. Thousand Oaks: SAGE Inc.

APPENDIX I - INTERVIEW GUIDE

Part I – Background on Lean Principles

Introduction

Mass production, an outgrowth of the industrial revolution, was presented in the US industry with the contributions of Henry Ford and F. W. Taylor in the early twentieth century, which formed the four main characteristics of mass production: reducing cost by increasing production volume, few innovations, supervised labors performing repetitive tasks and adversarial relationship with suppliers. Implementing mass production techniques enabled Ford to reduce production cost of the Ford Model T and sell the model at a price half of competition. However, the model lost its zest in the 1980s when it became apparent that the America auto industry is losing to foreign competition, particularly the Japanese. As a result, a commission lead by the Massachusetts Institute of Technology (MIT) was set up in 1986 to investigate the causes of the phenomenon. The results of the study were published in *The Machine That Changed The World*, a book by Womack, Jones and Roose which documented the superior performance of the Japanese Auto manufacturers in general, but Toyota company in particular based on *lean manufacturing*.

In a way, *lean* manufacturing is similar to mass production in a sense that it "creates outputs using less of every input". The difference between the two, however, is in the *lean* manufacturing's flexibility (ability to offer more variety to customers).

Since the early 1990s, *lean* manufacturing has been successfully implemented across several industries, ranging from small machine manufacturers, to the healthcare sector, and to even the aerospace industry. Lean principles, however, do not appear to have been applied in the Fast Moving Consumer Goods (FMCG) industries.

The aim of this research is therefore to examine the claimed universality by investigating whether *lean* is applicable to the Fast Moving Consumer Goods (FMCG) Industry.

The meaning of lean

When *The Machine That Changed The World* was published, Toyota's sales were half those of General Motors. In 2008, Toyota overtook General Motor's as the world's number one auto manufacturer. The book covers the practices of the Japanese auto industry in general, and the Toyota Production System (TPS) in particular. *Lean manufacturing* is the western name given to TPS.

Muda is the Japanese terminology for 'waste'. Value is the opposite of waste. Companies eliminate waste in order to create value, as perceived by the customers. Taiichi Ohno (1912 – 1990), the Toyota executive who masterminded TPS identified the original seven types of waste:

- *Overproduction.* Producing goods without having customer orders, which results in overstaffing and higher costs of storage and inventory. This is the "most serious" of the seven types of waste as it obstructs the smooth flow of product or information and inhibits quality and productivity.

- *Waiting.* Shop floor workers who set idle either due to wrong workload balancing, or just waiting for the next processing step, or because of out of stock of raw materials. It also occurs when goods are not moving or being worked on.

- *Unnecessary Transport.* The physical transportation of raw materials, finished goods or work in process (WIP). In addition to wasting time, this type of *muda* can also result in damage or deterioration of quality.

- *Over-processing.* Performing unnecessary processing due to poor product or process design. Over-processing leads to poor layout and results in further *muda* in the form of unnecessary transport and poor communication.

- *Excess Inventory.* In the form of raw materials, WIP or finished goods, which results in obsolescence, damage, and high transportation and storage costs. Inventory is a common way to hide problems (poor performance of suppliers or machines) and result in a significant increase in storage costs.

- *Unnecessary Movement.* Any wasted movement employees have to perform during their work, leading to poor productivity and possibility quality issues.

- *Defects.* The production of defective goods. In the TPS, each defect is seen as an opportunity for improvement rather than a problem.

Two additional types of waste were later added:

- *Poor Design.* Design of goods or services that do not meet customer needs.

- *Untapped employee potential.* Not using potential employee skills or ideas by not engaging them in the process.

In this sense, waste can be defined as "any human activity that absorbs resources but creates no value". Value, on the other hand, is the opposite of waste and is created as wasteful activities are reduced, or if additional features or services are added and are perceived to be valuable by the customer.

In TPS activities are categorized into three types (Liker, 2004, p. 280):

- *Value-adding activities [VA].* Activities that generate value in the eyes of the end customer of the product or service (Francis, 2004).

- *Non-value adding Activities [NVA].* Activities that clearly do not add any value to the product or service and can be removed in the short run with no or minimum capital investment (Francis, 2004). This is type referred to as Type Two *muda* (Womack & Jones, 2003, p. 20).

- *Necessary but non-value adding activities [NNVA].* Activities that do not add value but must be carried out due to constraints in technology, assets or procedures. This is type referred to as Type One *muda* (Womack & Jones, 2003, p. 20).

Lean tools and techniques

The following is a list of the most common *lean* tools, techniques and practices:

- *Value Stream Mapping Techniques.* This aims to distinguish value-adding form non-value adding activities. Several techniques can be used for this purpose: process activity mapping, supply chain response matrix, production-variety funnel, quality filter mapping, demand amplification mapping, value-analysis time profile and decision point analysis.

- *Continuous Improvement (Kaizen events).* One of the most basic yet powerful tools of *lean*. Once a problem is identified, *kaizen* events are used to find solutions to the problem. In these events, staff from various functions work together in order to find a solution.

- *Kaikaku.* Unlike *Kaizen*, this event involves major changes in the workplace including factory layout and machine rightsizing. It is typically carried out at the onset of any *lean* project.

- *Kanban.* Only produce to order, no more and no less, in order to create the flow of one.

- *Andon.* Designing operations to detect errors and stop the operation when an error occurs.

- *Poka Yoke.* Failure prevention by designing machines in an 'error proof' way.

- *Heijunka.* Production leveling by reducing batch size in order to make the best utilization of resources.

- *Takt-time.* The term was introduced in the German aircraft industry in the 1930s before it was later taken to Japan by German engineers training Japanese aircraft producers. Takt is a German word which means beat or rhythm. This tool is used to synchronize the production rate (or rhythm) with customer demand in order to avoid overproduction.

- *Jidoka.* This means automated stop system and is a function of machine design to allow machine to stop when an abnormal situation arises.

- *5S.* A visualization tool based on the conviction that organizing the workplace is essential to ensuring a smooth workflow. It is based on 5 Japanese words starting with the letter 's' which were later translated to equivalent English words:

  » *Seiri*: Sort,

  » *Seiton*: Set in Order,

  » *Seiso*: Shine,

  » *Seiketsu*: Standardize, and

  » *Shitsuke*: Sustain.

5S encourages workers to improve the physical setting of their work and teaches them to reduce waste, unplanned downtime, and in-process inventory. A typical 5S implementation would result in significant reductions in the square footage of space needed for existing operations. It also would result in the organization of tools and materials into labeled and color coded storage locations, as well as "kits" that contain just what is needed to perform a task.

REFERENCES

Francis, M. (2004) *Lean Information and New Product Introduction: An FMCG Case Study*, BNet [Online]. Available from: http://jobfunctions.bnet.com/abstract.aspx?docid=312599 (Accessed: 6 June, 2010).

Liker, J. K. (2004) *The Toyota Way:14 Management Principles From The World's Greatest Manufacturer*. New York: McGraw-Hill.

Womack, J. & Jones, D. (2003) *Lean Thinking*. New York: Free Press.

Part II - Interview Questions

We have just took a brief preview of *lean* principles, tools and techniques. The research aims to answer the following research question: How can *lean* principles be implemented in FMCG manufacturing and warehousing operations? Next I will be asking you a number of questions to get your views on the research subject. But first, I have this general question: in your view, would *lean* principles be applied in FMCG manufacturing and warehousing? Why do you believe so?

The first research objective is to determine whether or not lean principles can be applied in FMCG Manufacturing and Warehousing operations.

Q1 – Is *overproduction* a common practice in your industry?

– If the answer is yes: why do you think it is a common practice?

– If the answer is no: why do you think it is *not* a common practice?

Q2 – Are shop floor workers idle at any point of time during their working hours, for example: waiting for a machine to complete its cycle? Why do you think this does/does not happen?

Q3 – Are raw materials transported directly from their point of receipt to their point of consumption?

– If the answer is yes: Is there a reason why they are transported directly?

– If the answer is no: why is this the case? Would you consider the in-direct route a type of *unnecessary transportation*?

Q4 – During processing, are the semi-processed products transported directly from one machine to another?

– If the answer is yes: is there a reason why they are transported directly?

– If the answer is no: How are they transported? Why are they not transported directly?

Q5 – Once the production process is completed, are the finished products transported directly from the production line to the plant warehouse?

– If the answer is yes: Is there a reason why they are transported directly?

– If the answer is no: How are they transported? Why are they not transported directly?

Q6 – What is the average time a product spends in a plant warehouse before being shipped to a distribution center?

Q7 – Upon receiving goods in a distribution center, are they directly transported to the order picking area?

– If the answer is yes: Is there a reason why they are transported directly?

– If the answer is no: Why are they not transported directly?

Q8 – Upon picking an order in a distribution center, is the picked order directly loaded on the distribution vehicle?

– If the answer is yes: Is there a reason why they are directly loaded?

– If the answer is no: Why are they not directly loaded?

Q9 – Are there any layout design restrictions in your operations (production or warehouse) that results in *unnecessary processing*?

– If the answer is yes: What are they? Can they not be avoided? Why or why not?

– If the answer is no: go to the next question.

Q10 – Do you think that the inventory carried in your raw materials and finished goods warehouses are too high?

– If the answer is yes: Why do you need to keep such high inventories? What would be the impact of reducing these inventories? If the issues causing these high inventories are resolved, what do you think is the optimum level of inventory that you should be carrying?

– If the answer is no: Why do you think they are not too high? Are target inventory levels agreed with your finance department?

Q11 – In your opinion, do you think that the current machine layout in production (or the warehouse layout) is the most suited layout for maximizing the utilization of workers? If you are to redesign the production line (or the warehouse) layout from scratch, what would you change in order to minimize a) number of workers, b) the movement of workers?

Q12 – Let us quickly recap. From our discussion, we can say that of the seven types of waste, types [x, y & z] are most common in your industry. Taking this into consideration would you say that *lean thinking* can benefit your manufacturing/warehousing operations? Why or why not?

The second research objective is to examine the conditions under which lean principles can be applied in FMCG Manufacturing and Warehousing.

Q13 – Is it possible to start *lean* implementation in your operations under the current conditions?

– If the answer is yes: Why do you believe so?

– If the answer is no: Why? What are the conditions that must be changed before the implementation?

Q14 – Do you currently have a problem-solving/improvement system in place? Does problem solving involve management as well as floor shop staff?

Q15 – Would you consider your shop floor staff to be multi-tasked or specialized? Why do you believe so? Do you have a job rotation scheme in place?

Q16 – Do you think that the manufacturing/warehousing managers would be willing to support *lean* implementation in their operations? Why do you believe so?

Q17 – Do you think that the manufacturing/warehousing floor shop staff would be willing to support *lean* implementation in their operations? Why do you believe so?

Q18 – Is your current organization structure functional based or process/product based?

Q19 – Do you believe that your current organization structure is the right size for your operations? Why do you believe so?

The third research objective is to identify the lean tools and techniques that can be easily applied in FMCG Manufacturing and Warehousing.

Q20 – Briefly describe, in order of their occurrence, the main activities involved in your operations. Categorize each of these activities as being 'value-adding', 'non-value-adding' or 'necessary but non-value-adding'. Explain the reason for the categorization of each activity.

Q21 – Do you think *kaizen* events can improve your operations? Why or why not? How?

Q22 – Do you think a major *kaikaku* can help improve your operations? Why or why not? How?

Q23 – Do you think that using a pull *(Kanban)* system can improve your operations? Why or why not? How?

Q24 – Do you think that implementing an *andon* system can add value to your operations? Why or why not? Which areas can most benefit from it, if any? How?

Q25 – Do you think that implementing a *poke yoke* system can add value to your operations? Why or why not? Which areas could benefit the most from it? How?

Q26 – Do you think that using *takt-time* can add value to your operations? Why or why not? Which areas could benefit the most from it? How?

Q27 – In order for *lean* to be implemented a system of 'flow' must be created. This system, also known as just-in-time, calls for minimum – or if possible, zero – inventory in the process. From raw materials, to production, to plant warehouse, to distribution center to the customer. In order to be able to achieve this, the following infrastructure must be in place:

a. *Quick changeovers*: production must be done in as short runs as possible. This allows products to be produced only as needed. Longer production runs, although are efficient from the narrow local perspective of production, will generate excess inventory and increase operational cost. In order for the short runs to be possible, the changeover between any two products must be minimal.

b. *Production leveling*: fluctuations in demand, or sudden increases or drops, make production planning more difficult for the plants as well as for the raw materials suppliers. The ability to level production across months, weeks and days with minimum variation will provide the stability required to achieve the flow.

- Do you think that the current changeover times are reasonable? Can they be reduced from their current level to be less than 30 minutes? If yes, how? If no, why?

- If changeover times are actually reduced to be 30 minutes or less. Are there other obstacles that discourage the concept of short production runs?

Q28 – Do you think that implementing the *Heijunka* (production leveling) principle is beneficial in the manufacturing/warehousing operations in your industry?

- To be able to level production across seasons, is it possible to give customer/consumer promotions during low seasons to encourage higher sales during low seasons and lower sales during peak months? If no then why?

- Are sales normally higher towards the end of the month? If yes, why? Is it the consumer pick up trend? Or is it a sales trend? Can this be leveled? If yes, how? If no, why?

Q29 – Do you think that implementing a *jidoka* system can add value to your operations? Why or why not? Which areas can most benefit from it, if any? How?

Q30 – Do you think that implementing the 5S system can add value to your operations? Why or why not? Which areas can most benefit from it, if any? How?

The fourth research objective is to look at the benefits and drawbacks of applying lean principles in FMCG Manufacturing and Warehousing.

Q31 – One of *lean* basics is to ensure that an organization adds value to its customers through their operations and through every new product developed. Do you believe that the 'voice of the customer/consumer' is heard in your organization? Why do you believe so?

Q32 – How is quality measured in your industry? Do you have a high percentage of defects in your operations (either defective goods, or wrongly processed orders)?

- If the answer is yes: Why? Can they not be avoided? Why or why not?

- If the answer is no: Why do you think that the percentage is not high?

Q33 – How is operational cost measured in your operations? How is the cost of defects measured in your operations?

Q34 – With reference to our discussion: in general, do you think that:

- implementing lean principles and using lean tools and techniques will benefit your operations? Why? In what ways?

- Are there any drawbacks to implementing lean principles and lean tools and techniques in your operations? What are they?

AUTHOR'S BIOGRAPHY

Alaa Aljunaidi: Is currently the Director of Logistics & Lean Developments SIBCO, a PepsiCo bottler in Saudi Arabia. He completed his B.Sc. in Mechanical Engineering from the American University in Cairo, Egypt and his B.Sc. in Operations and Supply Chain Management from the University of Liverpool, UK. This research was completed for his dissertation in this program.

Samuel Ankrah: Is Co-Chief Executive of Specialist Health Ghana, a company involved in Management Consultancy and Heath Related Products. He has over twenty-nine years of management experience, several years of experience in technology and innovations and is Honorary Lecturer for University of Liverpool and University of Roehampton, both in the United Kingdom.

7

Operations in the context of urban mobility: Evaluating the Performance of the Deliveries to Small Retailers

João Victor Rodrigues Silva
martins@cepead.face.ufmg.br
Center for Graduate Studies and Research in Business Administration (CEPEAD)/Federal University of Minas Gerais (UFMG)

Ricardo Silveira Martins
martins@cepead.face.ufmg.br
UFMG

ABSTRACT: This study aims to clarify the performance attributes of logistical service that retailers identify as more relevant to their suppliers' distribution process in urban areas. And, this task was performed using the Stated Preference Technique, a tool not so widespread among operations management researchers. The research focused the small retail food stores in Belo Horizonte, Brazil, considering the complexity involved in the delivery of perishable and fragile foods. As result, the technique show us the relative importance of the attributes related to the service level, highlighting the high degree of relevance achieved by the attributes that define the service level. This means that gaps in services are not easily shadowed by compensatory practices of stock build-up, which are financially very burdensome or unviable in terms of storage area. This kind of approach may call attention to opportunities of improving buyer-supplier relationships and the management of these processes, as well as public issues.

Key words: Small retailer, Urban distribution, Supply performance, Stockout.

1. INTRODUCTION

The urban concentration creates problems related to the urban mobility of people and decline in the quality of life, and hinders the supply of commercial establishments. In the similar context, the delivery process related to the handling of products also poses some challenges. The effects can be perceived in different ways, from increased costs to quality of services. The increase in costs can be attributed to a number of factors like, the need for an expanded fleet, smaller orders in more frequent batches, and the formation of a higher level of safety stock. Meanwhile, the quality of service may fall, for example, measured by the number of deliveries delayed.

Beyond that, as the problem of mobility in urban centers worsens, these effects become more harmful for customers and retailers. To the end customer, the most perceived results are an increase in prices and a decrease in the availability and variety of products on the shelves. For the retailer, in particular, the level of service, the rising costs, and the customers' reactions to such situations results in a loss of sales and a decline in business profitability (Ali 2011; Aastrup & Kotzab 2010).

Retail companies, on daily basis, experience complex situations, involving challenges to provide their customers with a level of service that is consistent with expectations, while at the same time, at controlled costs. This requires a greater capacity in logistical planning, along with efforts in the traditional areas, such as marketing, finance, and customer service. Planning, executing, monitoring, measuring, and controlling the performance of logistical operations in large urban centers is the key for success in the food retail segment as it involves direct contact with the end consumer.

Moreover, it is customary for suppliers to lack clear information regarding the actual needs and wants of retailers. Some authors not only demonstrate that customers are interested in more than just costs, but also highlight the need to investigate other attributes linked to service (Ballou 2004; Bowersox, Closs & Cooper 2002; Chopra & Meindl 2007).

Therefore, this study aims to clarify the performance attributes of logistical service that retail businesses identify as more relevant to their suppliers' distribution process in urban areas. And, this task was performed using the Stated Preference Technique, a tool useful for leading with this kind of phenomenon, since we can analyze scenarios devised by the respondents so that they make the trade-offs between attributes of service, but not so widespread among operations management researchers.

To deal with this question, it is first necessary to create mechanisms that would help identify which are the needs of retailers in relation to their partners, so that consumers receive higher added value. Therefore, the knowledge of logistical performance indicators is an important tool in this investigation (Phusavat et al, 2009; Barros et al, 2010).

2. PERFORMANCE INDICATORS IN TRANSPORT AND OPERATIONS

Urban transportation is essential in the context of the modern economy. Plowden and Buchan (1995) show that an efficient urban transport system should offer goods or services to customers at a reasonable cost.

Measures to improve cargo handling in urban areas involve the sharing, participation, and cooperation of shippers, suppliers, transporters, retailers, and customers in the implementation of activities. Therefore, it is necessary to use technologies that align with the reorganization of the entire supply operational processes and organizational elements (Anderson, Allen, & Browne 2005; Qureshi, Kumar & Kumar 2008). The use of performance indicators for operation has proven to be a helpful organizational element in overcoming the difficulties in distribution logistics and a good mechanism for business improvement.

Companies already exploit the earning opportunities provided by logistics. However, the evaluation of performance using metrics and indicators is still in the embryonic stage, specifically when the focus is transactions between companies (Gunasekaran, Patel & Tirtiroglu 2001; Neely, Gregory & Platts 2005). Without these measures, organizations struggle to analyze whether their strategies are consistent with previously established objectives (Phusavat et al, 2009) and to propose measures that address potential problems with their feasibility (Gunasekaran, Patel & Tirtiroglu 2001; Barros et al, 2010). Based on a greater understanding of performance metrics, they may review the organizational objectives and re-engineer processes (Neely et al, 2000).

In the case of urban freight transport, Gunasekaran, Patel and Tirtiroglu (2001) relate a number of upstream (suppliers, shippers, wholesalers) as well as downstream (distributors, retailers) delivery factors in the supply chain, such as vehicle speed, driver reliability, frequency of deliveries, and location of deposits with the formation of stock.

In addition, a failure to obtain an optimal location may be considered as a factor if the total cost of the process is less due to gains in other departments, areas, or activities. According to Chen (2008), the total cost approach is most useful in evaluating the performance of distribution activity by including non-financial measures that are aligned to the objectives of the firm and oriented towards customers.

Based on the literature review and recommendations of authors, like Neely, Gregory and Platts (2005) and Slack, Chambers and Johnston (2007), the measurement system of this study will use the performance indicators derived from the following operational attributes: Quality, Speed/promptness, Reliability/consistency, Flexibility and Cost.

As a performance attribute, quality is defined as responsibility to exceed the satisfaction expected by the customer (Neely, Gregory & Platts, 2005). Promptness in the distribution process is a source of competitive advantage and an important performance attribute. Changes in the logistical process are undesirable and should be addressed in order to achieve greater consistency and reliability with the agreed deadlines (Goldratt & Cox, 1986). Reliability or consistency, as defined by Slack, Chambers and Johnston (2007), concerns the responsiveness of a company or supply chain in meeting the customer's requirements of confidence in the logistical process.

Flexibility includes the dimension of variety, which is consistent with the ability of a logistical system to be quickly modified to meet a need (Slack, Chambers & Johnston 2007). Finally, cost is always important in any situation, especially in the urban cargo transport, which is always under pressure due to this attribute.

Based upon the defined importance of performance attributes, we now describe the methodology used to create a system of performance measurement for the urban distribution to small retail food stores in Brazil.

3. METHODOLOGY

This is characteristically an inductive study, using both qualitative and quantitative research methods.

Another methodological feature of this study is that it is exploratory in nature, and aims to elucidate those performance attributes that are most relevant to the retail food sector in Belo Horizonte (5 million inhabitants in metropolitan area), Brazil.

The data was collected using the survey method to enable a better understanding of the phenomenon. This research intends to create a model of performance measurement based on the scores of indicators and attributes assigned by a group of respondents to generalize the behavior of other members who may not have participated in the survey.

The unit of analysis defined for this study are small retail food stores, more precisely bakeries, grocery stores, and small supermarkets, in the urban area of Belo Horizonte, justified by the guiding question of the work itself. Belo Horizonte's retail sector is engulfed within the urban environment of deteriorating mobility and faces challenges that fall the performance of logistics distribution.

Additionally, the end consumer is the last bond. There is an increasing responsibility to provide services desired by consumers due to the associated threat of customer dissatisfaction, lost sales, decline in revenue, as well as business risk, particularly in the food sector (Ali, 2011).

This study consists of two phases. In the first phase, we administered a semi-structured interview to 15 retail managers in Belo Horizonte, Brazil, based on the stated preference technique. We selected the sample based on researcher intentionality and respondent accessibility. Next, we obtained the distribution management reports of the metropolitan region of Belo Horizonte, and evaluated the retail performance attributes, namely, *cost, service, delivery, reliability, and flexibility.*

3.1 Research steps and methods

The semi-structured interview was administered to 15 managers and entrepreneurs in the urban retail food sector to identify their vision of the logistical distribution of their suppliers in the urban environment, as well as to understand the factors that they consider critical in this process and the measures that need to be taken to improve their satisfaction.

Next, stated preference cards were applied. Li and Hensher (2012) maintain that stated preference is a method to reveal preferences and behaviors based on attributes and multiple levels. To identify the relative importance of each characteristic or attribute under evaluation, stated preference enables organizations to adjust their products or services to the customers' wishes. In this case, too, the intent was to capture the ranking of the service attributes through four rounds of interviews of an independent nature, as presumed in the literature. In this manner, these four rounds of interviews yielded 60 independent responses.

The attributes adopted for this research, as proposed by Slack, Chambers and Johnston (2007), are derived from the model that measures the operational performance of urban food retail stores with regard to the logistical distribution of suppliers, as explained in Table 1.

The assembly of the cards for the presentation of alternatives to the respondents depends on the number of attributes and the options available to choose from each attribute. As Figure 1 indicates, this research adopted five attributes and two alternatives for each attribute.

Figure 1 Attributes, levels, and numerical codes used in the Stated Preference

ATTRIBUTE	LEVELS	NUMERICAL CODE
DELIVERY	Quick: corresponds to shorter order cycles (the period between the execution of the order and the shipment of the items or the start-up of the service). It implies faster reactions to the surprises offered by the demand.	1
	Slow: longer order cycles are not desirable for agencies and entities, since this forces them to make a detailed plan for the demand, which is not always possible due to the emergency nature of some requests (for example, disease outbreaks, riots).	0
SERVICE	Satisfactory: customer service is considered effective in terms of both the information provided and the products delivered or services performed according to specifications previously made available.	1
	Low: customer service is hampered by the inaccuracy of information, service errors, and contradictions between the contractor and that which has effectively been provided.	0
RELIABILITY	Accuracy of service: customers are certain when they will have their demands met. The agreed time limits are strictly adhered to.	1
	Imprecision of service: users have some level of uncertainty as to when their demands will be met.	0
FLEXIBILITY	Possibility of changes: the mix of products and services offered, as well as set times for service, can be changed upon customer's demand.	1
	Changes are not allowed: modifications related to the mix offered, either in terms of product/service or service hours.	0
COST	Transactions made at a higher cost to the customer.	0
	Lower cost for distributing products and services.	1

In this study, the stated preference technique was used to define the relevant attributes of logistical performance indicators in the urban retail food sector in Belo Horizonte, as well as to highlight the relative importance of each attribute for the urban retail food sector. The process of choosing the cards was a ranking of the alternatives based on the utility of the retail managers.

The next step, following the validation and relativization of the attributes' importance, consisted in the selection of indicators that, in fact, will compose the system of performance measurement. For this research, we used literature review to pre-select 26 indicators related to the attributes *cost, service, delivery, reliability, and flexibility*.

At first, the same retailers who validated and identified the most important attributes for urban logistical supply operations with regard to their establishments were asked to choose the performance indicators to form the assessment system. However, during the implementation of the stated preference technique, these managers found it difficult to select and identify the function of each indicator. This happened when the performance indicators were shown to the respondents, and they were asked to suggest, criticize, or even exclude an indicator. Perhaps, they have struggled to prioritize the performance indicators of operations to business needs.

The indicators were then presented to the body of experts, Masters and Doctoral students in the Disciplines of Operations of the Graduate Program in Administration, at the Federal University of Minas Gerais, and logistics professionals. They ranked the indicators in the order of priority, assigning a grade of 5 for indicators of high priority, a grade of 3 for those of medium priority, and a grade of 1 for indicators of low priority.

According to Gunasekaran, Patel and Tirtiroglu (2001) and Neely, Gregory and Platts (2005), the number of indicators is an extremely important factor for managers, since excess measures can take the focus away from what really impacts the operation, and should be monitored and evaluated. Thus, 11 indicators were selected to form the performance measurement system for the distribution of food to the micro and small retail sector of Belo Horizonte.

The classification of the indicators followed the ABC criterion, whereby a few items (indicators) reflect the greatest number of cases. Thus, the respondents

chose and classified the indicators as follows, as used by Coelho, Martins and Lobo (2013):

- 20% of the indicators as high priority;
- 30% of the indicators as medium priority; and
- 50% of the indicators as low priority.

After the application of indicator questionnaire and achievement of the results, we followed a new classification, based on the sum of the grades for each indicator. At this time, we made a new selection according to the established criteria in order to define the indicators that will compose the measurement system of logistical performance for urban distribution. Finally, we established the performance indicators, and balanced them according to the weight of each defined attribute at the stated preference technique application step.

3.2 Data analysis

To analyze the data resulting from the stated preference technique, the multinomial logit model was used. Assuming that the error term of the utility function is governed by a Gumbel distribution, the multinomial logit (Ben-Akiva & Lerman, 1985) can be written as follows:

$$P_n(i) = \frac{e^{\beta_k X_{ink}}}{\sum_{j \in C} e^{\beta_k X_{jnk}}}$$

in which $P_n(i)$ is the probability of the alternative *i* being chosen by the individual *n* within a construct of possibilities *C*.

Statistical responses were used for the analysis of the responses by the algorithm developed by Souza (1999).

4 RESULTS

4.1 Validation of attributes and definition of weights

The technique of application of stated preference cards aims at validating the choice of attributes and defining the relative importance of each attribute in relation to the others to approximate the choice of indicators and the format of the performance measurement system.

In implementing the experiment, it was decided not to present all possible combinations of levels and constructs to respondents, that is, to use the fractional factorial. The obtained results are derived from the ranking by respondents in six groups of five cards, with eight cards using the technique of partially-balanced incomplete groups suggested by Souza (1999). Often, a single repetition of a factorial experiment goes beyond the capabilities of researchers, or provides more precision than needed for the estimation of the main effects (Cochran and Cox 1978).

The use of fractional repeat experiments was proposed by Finney (1945). Since then, these experiments have been used in many applications, especially in industrial development. Their main attraction is that they allow the inclusion of five or more factors in an experiment of practical size, so that the researcher can quickly determine the effects of the factors in the outcome.

The parameter estimation was based on the multinomial logit model with conditional probability (Table 1). As a premise, the multinomial logit model defines the preferences of the retailers as homogeneous. The estimators of the attributes represent the marginal utility of each attribute and provides it with an equal meaning (Marcucci & Gatta, 2012).

Table 1 Stated preference result LMPC model

Attribute	Coefficient	Error	T test	IC (t=25%)
Flexibility	0.1784	0.2309	0.7728	[-0.283; 0.640]
Delivery	1.1426	0.2565	4.4540	[0.630; 1.656]
Cost	1.4939	0.2478	6.0286	[0.998; 1.990]
Service	1.4658	0.2407	6.0910	[0.985; 1.947]
Reliability	0.9695	0.2423	4.0013	[0.485; 1.454]

Number of interviews = 60   Number of cases = 180

Rho= 0.3486   Rho (Ajt) = 0.3224

The analysis of the results reveals that the *cost* attribute displayed the greatest relative importance among the five selected attributes, with a coefficient of 1.4939. This means that bakeries, grocery stores, and small supermarkets in the sample perceive the predominant importance of the *cost* attribute. The *service* attribute was second in level of importance to those surveyed, with a parameter of 1.4658, very close to the *cost* parameter. Next, among the attributes of the operation of the urban distribution of food, *delivery* (1.1426) and *reliability* (0.9695) were most important. The least important among the five attributes was *flexibility* in the logistical operation.

Based on the criterion of representativeness, which considers the weight of each attribute in relation to the total value of the coefficients, the *cost* attribute obtained 28.45% of the total weight of the coefficients. This was followed by the *service* attribute, which showed a relative weight of 27.92%, the *delivery* attribute at 21.76%, *reliability* at 18.47%, and *flexibility* at 3.40%.

The next step analyzed the significance of each parameter or attribute according to a model. The LMPC software (Souza, 1999) used considers the *t*-test with a significance level of 95% for these parameters. To perform this analysis, Hair et al. (2005) argue that a good measure of significance for the t-test is to consider values greater than one standard deviation, or 1.96. Therefore, attributes above this measure are statistically different from 0. The attributes with a t measurement below 1.96 are statistically equal to 0.

All the parameters under consideration differed significantly from zero, with the exception of *flexibility*. This claim that they are different from zero can be made with 95% certainty [*delivery*, with a t-test of (4.4540), *cost* (6.0286), *service* (6.0910), and *reliability* (4.0013)]. However, *flexibility* in the logistical distribution of food products in the urban area of Belo Horizonte could not be considered as an attribute significantly different from zero with 95% certainty, as its t-statistic score is 0.7728. Therefore, this attribute was removed from the elaboration of the performance measurement system.

The reliability construct has a higher correlation coefficient, indicating a very strong correlation, which means that the population sampled places a lot of importance upon this attribute, and that the attribute of flexibility has less importance statistically because its correlation coefficient is very weak. According to Shikimura (2006), a very strong association exists at values of 0.90 and 1.00 and a very weak correlation at values between 0.00 and 0.19.

According to the approach taken in the urban freight transport literature, cost still emerged as the main attribute in the formation of urban retailers' expectations, and therefore, a key point in planning a logistical operation or a city's urban roads. However, the results of this study indicate that, at least with regard to the micro and small retail food segment in the capital of Minas Gerais, cost should not be considered as the only important factor for evaluating the performance of a distribution operation in an urban environment.

This result corroborates with Morgan and Hunt (1994), Johnson and Grayson (2000), and Lindgreen (2003), who highlight the importance of intra- and inter-organizational relationships along the supply chain in search of cooperation, credibility, and trust among partners to increase the efficiency of the operation or process, and improve the profitability of the entire chain. These studies consider that the gains, thus achieved, are measured in terms of the reduction of opportunistic behavior, the elimination of predatory bargaining power, and the reduction of transaction costs and the switching of vendors.

Together, the relative importance of the *service, reliability,* and *delivery* attributes in the operation of retail supply was around 70%. The result may be explained by the impact of the lack of products on the shelves (stockout) on retail performance (Woensel et al, 2007; Aastrup & Kotzab, 2009, 2010).

4.2 Definition of performance indicators

We originally utilized 26 indicators obtained from Bowersox, Closs and Cooper (2002), Ballou (2004), Slack, Chambers and Johnston (2007) and Neely, Gregory and Platts (2005).

However, as the application of the stated preference technique revealed no significance for the *flexibility* attribute in the evaluation model of the proposed logistical performance, the final list comprised 22 indicators.

Fifteen of the 18 experts provided responses, representing 80% of the experts' population. The indicator was graded at 5 if the expert defined it as a high priority, 3 if defined a medium priority, and 1 if defined a low priority.

Table 2 describes the set of indicators defined for the system.

Table 2 Indicators proposed for the performance measuring system of the food distribution operation in the retail sector of Belo Horizonte

ATTRIBUTE	INDICATOR	DEFINITION	FÓRMULA
SERVICE	Product /service according to specifications	Product delivered or service performed according to specifications.	Semantic differential scale of six points – extremes: from fully compliant to divergent.
	Service quality (agile confirmation, friendliness, promptness)	Customer perception of service quality (from electronic to personal contact)	Semantic differential scale of six points – extremes: from poor to optimal.
	Returns	Percentage of returned products or unapproved services within a proposed period to the total number of products delivered and services performed during the period.	Sum of the number of returned products or unapproved services by the customer/Total number of products delivered and services performed during the period.
	Complete orders	Percentage of orders delivered whose requests were fully met (mix and quantity) for the number of orders placed during the period.	Number of complete orders filled/Total number of orders placed during the period.
RELIABILITY	Product integrity	Percentage of products delivered undamaged in relation to the total number of products delivered in the period.	Number of products delivered undamaged/Total number of products delivered in the period.
	Complaints regarding orders	Percentage of orders delivered whose requests were not fully met (assortment or quantity) for the number of orders placed during the period.	Number of orders serviced with a complaint/Total number of orders placed during the period.
COST	Percentage of the transport cost in sales	Share of total transport costs in relation to sales.	Total shipping cost/revenue earned through sales.
	Cost of out-of-stock products for sale	Measurement of the loss of profitability by the absence of inventory to meet demand.	Semantic differential scale of six points – extremes: from no loss to significant loss.
DELIVERY	Average length of order delay	Average length of delay after the deadline established for the order's submission.	(Time in day and hour of receipt of the product or initiation of service) – (time in day and hour previously established for order fulfillment)
	Length of the order cycle	Length of time between placing the order and its actual delivery.	(Day and hour of submission of the order) – (Day and hour of order placement by the customer)
	Waiting time for receipt of the complaint	Time elapsed since the identification of a delivery complaint until its correction (complementation).	(Time in day and hour of the receipt of the item of complaint) – (time in day and hour when the complaint was identified)

5. FINAL CONSIDERATIONS

The issue of urban mobility motivated this study, which aims to define the most important attributes to evaluate the performance of the logistical operation for retail supply.

Urban mobility in large urban centers has decreased considerably in recent years, deteriorating the quality of life and level of logistical supply service to companies in those areas. In this framework, we define the small retail food stores in Belo Horizonte, Brazil, as the unit of analysis for the proposal to develop a system to evaluate the performance of suppliers and carriers in the logistical operation of supply in these stores. The study focuses on the logistical operation of food supply due to the complexity involved in the delivery of perishable and fragile foods. Thus, quality in this type of operation is critical and measurement of the performance of activities becomes essential for companies included in this supply chain.

The performance measurement system enables the measurement and control of the activities. Moreover, its systematic use would result in a cycle of continuous improvement in the food supply chain, thereby providing a higher quality of service to retail stores, and consequently, to the end consumer. Furthermore, the proposed system helps identify problems and can be seen as a tool for integration.

Another noteworthy point of this study is the determination of importance of the attributes related to the service level. The high degree of relevance achieved by the attributes that define the service level may indicate bottlenecks and current critical points in services. This means that gaps in services, such as delays in delivery or failure to deliver completed items of an order, compromise the availability of products on the shelves and this situations of stockout has serious implications for the retail store in relation to the customer. At the same time, compensatory practices of stock build-up are financially very burdensome or unviable in terms of storage area.

As a limitation of this study, it can be understood that the retail segment under study has some peculiarities. Furthermore, the experiment was conducted in Belo Horizonte, Brazil, which also has its own unique qualities.

However, despite these limitations, it is evident that the model's gradual development tests logistics, while also contributing to the suggestion of the use of Stated Preference for more assertive positions

concerning attributes and their relative importance. Thus, this study contributes to business management by developing a tool for control and performance monitoring. Its constant use will enable the creation of a cycle of continuous improvement, with the establishment of goals, research on problems, and proposal of solutions.

Another possible application of the proposed system is for boards of management and regulation as an indicator of urban mobility, because retailer satisfaction with the operation of their suppliers and carriers is linked to traffic conditions and cargo handling in urban centers (Kohler, 1997; Taniguchi, Heijden & Van Der, 2000; Kohler, 2001, Taniguchi, Thompson & Yamada, 2001, Taniguchi et al, 2001; Stewart, 1995; Hsu, Kannan & Leong, 2008).

6. REFERENCES

AASTRUP, J.; KOTZAB, H. (2009) Analyzing out-of-stock in independent grocery stores: an empirical study. **International Journal of Retail & Distribution Management**. 37:765-789.

_____. (2010) Forty years of out-of-stock research – and the shelves are still empty. **The International Review of Retail, Distribution and Consumer Research**. 20:147-164.

ALI, S. S. (2011) Redefining retailers satisfaction Index: A Case Of Nestle India Ltd. In: 22nd Annual Conference of Production & Operation Management Society, 2011, Peppermill Resort Spa Casino, Reno, Nevada, U.S.A. **Proceedings...** Peppermill Resort Spa Casino, Reno, Nevada, U.S.A.

ANDERSON, S.; ALLEN, J.; BROWNE, M. (2005) Urban logistics - how can it meet policy makers' sustainability objectives? **Journal of Transport Geography**, 13: 71-81.

BALLOU, R. H. (2004) **Business Logistics**: Supply Chain Management. 5th Edition, Upper Saddle River, New Jersey: Pearson-_Prentice Hall_.

BARROS, A. C._et al._ (2010) A framework for evaluating firm-level supply chain performance. In: 17th International Annual EurOMA Conference. Proceedings…, Porto, Portugal.

BEN-AKIVA, M., LERMAN, S. R. (1985) **Discrete Choice Analysis:** Theory and application to travel demand. New Jersey, Mit Press.

BOWERSOX, D. J.; CLOSS, D. J.; COOPER, M. B. (2002) **Supply Chain Logistics Management**. McGraw-Hill, New York. (Series Operations and Decision Sciences)

CHEN, C. C. (2008) model for customer-focused objective-based performance evaluation of logistics service providers. **Asia Pacific Journal of Marketing and Logistics,** 20: 309-322.

CHOPRA, S.; MEINDL, P. (2007) **Supply Chain Management**: Strategy, Planning, and Operations. 3rd ed., Upper Saddle River, New Jersey: Pearson Education, Inc.

COCHRAN, William G.; COX, Gerturude M. **Diseños Experimentales**. México : Trillas, 1978

COELHO, R. R; MARTINS, R.S.; LOBO, D.S. (2013) Desenvolvimento de Modelo de Avaliação de Desempenho: Aplicação a um Centro de Serviços Compartilhados. **Revista Portuguesa e Brasileira de Gestão**, 12: 69-80.

FINNEY, D.J. (1945) The fractional replication of factorial arrangement". **Ann. Eugen.** 12, p. 291-301.

GOLDRATT, E. M.; COX, J. (1986) **The goal**: beating the competition. Hounslow.

GUNASEKARAN, A.; PATEL, C.; TIRTIROGLU, E. (2001) Performance measures and metrics in a supply chain environment. **International Journal of Operations & Production Management**, 21: 71-87.

HAIR, J.F.; ANDERSON, R.E.; TATHAM, R.L.; BLACK, W. C. (2005) **Multivariate Data Analysis**, Upper Saddle River, NJ: Prentice Hall.

HSU, C.; KANNAN, V.; LEONG, K. (2008) Information sharing, buyer-supplier relationships, and firm performance. **International Journal of Physical Distribution & Logistics Management**, 38: 296-310.

JOHNSON, D.; GRAYSON, K. (2000) Sources and Dimensions of Trust in Service Relationships. IACOBUCCI, D.; SWARTZ, T. **Handbook of Services Marketing**, Thousand Oaks, CA: Sage, 357-370.

KOHLER, U. (1997) An innovating concept for city-logistics. In: Proceedings 4th World Congress on Intelligent Transport Systems, 1997, Berlin, Germany. **Proceedings...** Berlin, Germany.

KOHLER, U. (2001) How to change urban traffic with city logistics. In: Schneider, E., Becker, U. (Ed.). **Control in transportation systems**, Pergamon, Oxford, p.199-201.

LI, Z.; HENSHER, D. A. (2012) Congestion charging and car use: a review of stated preference and opinion studies and market monitoring evidence. **Transport Policy**, 20: 47-61.

LINDGREEN, A. (2003) Trust as valuable strategic variable in the food industry: Different types of trust and their implementation. **British Food Journal**, 105: 310-327.

LOUVIERE, J. J.; HENSHER, D. A.; SWIAT, J. D. (2000). **Stated Choice Methods**: analysis and application. 1st ed. Cambridge: Cambridge University Press.

MARCUCCI, E.; GATTA, V. (2012) Dissecting preference heterogeneity in consumer stated choices. Transportation Research Part E: **Logistics and Transportation Review**, 48: 331-339.

MORGAN, R.M.; HUNT, S.D. (1994) The commitment-trust theory of relationship marketing. **Journal of Marketing**, 58: 20-38.

NEELY, A.; GREGORY, M.; PLATTS, K. (2005) Performance measurement system design: a literature review and research agenda. **International Journal of Operations & Production Management**, 25:1.228-63.

NEELY, A. *et al.* (2000) Performance measurement system design: developing and testing a process-based approach. **International Journal of Operations & Production Management**, 20: 1.119-45.

ORTÚZAR, J. D. (1998). **Modelos de demanda de transporte**. 2nd ed. Santiago: Ediciones Universidad Católica de Chile.

PHUSAVAT, K. *et al.* (2009) Performance measurement: roles and challenges. **Industrial Management & Data Systems**, 109: 646-64.

PLOWDEN, S.; BUCHAN, K. (1995) **A new framework for freight transport.** London.

QURESHI, M. N.; KUMAR, D.; KUMAR, P. (2008) An integrated model to identify and classify the key criteria and their role in the assessment of 3PL services providers. **Asia Pacific Journal of Marketing and Logistics**, 20: 227-249.

SLACK, N.; CHAMBERS, S.; JOHNSTON, R. (2007) **Operations Management**. 5th Ed. Harlow: FT Prentice Hall.

SOUZA, O. A. (1999) Delineamento experimental em ensaios fatoriais utilizados em preferência declarada, Florianópolis, UFSC. (Tese - Doutorado em Engenharia de Produção/Universidade Federal de Santa Catarina)

STEWART, G. (1995) Supply chain performance benchmarking study reveals keys to supply chain excellence. **Logistics Information Management**, 8: 38-44.

TANIGUCHI, E.; HEIJDEN, R. E.; VAN DER, C. M. (2000) An evaluation methodology for city logistics. **Transport Reviews**, 20: 65-90.

TANIGUCHI, E.; THOMPSON, R. G.; YAMADA, T. (2001) Recent advances in modelling city logistics. In: TANIGUCHI, E.;THOMPSON, R. G. *et al* (Ed.). **City Logistics II**.. Kyoto: Institute of Systems Science Research. pp. 3-33.

TANIGUCHI, E. *et al.* (2001) **City Logistics**: network modelling and intelligent transport systems. Oxford: Pergamon, 2001.

WOENSEL, T.; DONSELAAR, K., BROEKMEULEN, R.; FRANSOO, J. (2007) Consumer responses to shelf out-of-stocks of perishable products, **International Journal of Physical Distribution & Logistics Management**, 37: 704-718.

AUTHOR'S BIOGRAPHY:

João Victor Rodrigues Silva: is an Economist and received his Master Degree in Administration from the Center for Graduate Studies and Research in Business Administration, Federal University of Minas Gerais (UFMG), Brazil. His research interests include urban distribution, transport systems and Regional development and transportation.

Ricardo Silveira Martins: Doctor in Applied Economics from the University of São Paulo. Lecturer at UFMG. Teaches and carries out research in Operations and Logistics Management at the Post-Graduate Centre for Research in Management. One of the leaders of the Interdisciplinary Centre for Research in Logistics. Areas of activity: Supply Chain Logistics, Logistics in Small and Medium-sized Companies and Logistic Strategies.

8

An Integrated Decision Model for Selection of Third Party Recovery Facilitator (3PRF) for Product Recovery Operations

Divya Choudhary
divyachoudhary2626@gmail.com
Indian Institute of Technology

Jitendra Madaan
jmadaaniitd@gmail.com
Indian Institute of Technology

Rakesh Narain
narainr@mnnit.ac.in
Motilal Nehru National Institute of Technology

ABSTRACT: Due to intricacies, considerable risks are involved in product recovery operations, therefore core competency and experience are prerequisite for successful implementation of Reverse Logistics (RL). To overcome this, organizations are outsourcing the recovery operations and are opting for Third Party Recovery Facilitators (3PRF) to obtain prominent results. This paper contributes by proposing a decision model to address risks associated with product recovery and suggesting criteria for suitable 3PRF selection. The study also introduces a novel approach, combining Interpretive Structural Modeling (ISM) and ELECTRE III methodologies to tackle this problem of 3PRF selection to improve decisions. ELECTRE III is a more reliable and pragmatic method, as it considers imprecision present in expert opinions and ensures that an extremely low score on a criterion cannot be compensated by good scores on other criteria. Subsequently, ISM methodology is used for identifying relationships among various criteria to determine the most significant ones. ELECTRE III method is then applied to select the most appropriate 3PRF based on the most significant criteria. The application of this novel approach can reduce risk involved and time during the process of recovery facilitator selection. Additionally, case study has been conducted to show the practical implementation of ELECTRE III methodology in selection process.

Keywords: Third Party Recovery Facilitators (3PRF), Reverse Logistics (RL), Interpretive Structural Modeling (ISM), ELECTRE III, Decision Model.

1. INTRODUCTION

Presently, consumption of virgin resources is increasing rapidly worldwide, as a result of escalating population and requirements of people. Accordingly, reckless consumption of natural resources is having a critical impact on our environment and is leading to unsustainable development of our society. Therefore, to restrain this uncontrolled exploitation of our ecological resources, concept of product recovery is being promoted across the world. This scenario has resulted in emergence of a new practice called Reverse Logistics (RL), which is a process of reclaiming value from end-of-use, end-of-life, obsolete and commercial returns with the help of various recovery operations, namely, recycle, resale, refurbish, cannibalization, remanufacture, repair and disposal. Further, according to Keong, 2008, 240,000 tons of virgin resources can be saved, if all the people across the globe return just one mobile device at the end of its life. This can curb emission of greenhouse gases to same extent as removing 4 million cars off the road. So, it can be affirmed that RL has enormous environmental scope and can contribute towards sustainable development.

Reverse logistics is quite opposite of forward logistics as it involves backflow of returned products from customers to manufacturers. Accordingly, logistics is defined as " the process of planning, implementing, and controlling the efficient, cost effective flow of raw materials, in-process inventory, finished goods and related information from the point of origin to the point of consumption for the purpose of conforming to customer requirements"(Council of Logistics Management, 2003). On the other hand, RL can be defined as "The process of planning, implementing, and controlling the efficient, cost effective flow of raw materials, in-process inventory, finished goods, and related information from the point of consumption to the point of origin for the purpose of recapturing value or proper disposal" (Rogers and Tibben-Lembke,1999). The process of reverse logistics can be depicted as shown in figure 1 given below:

Figure 1- Reverse Logistics Process

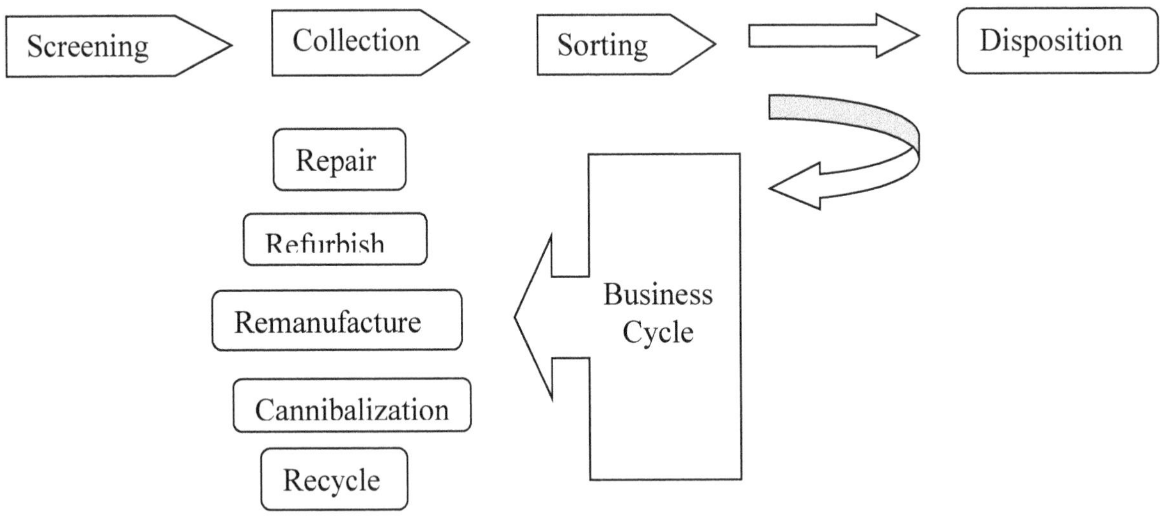

The first phase of RL is screening, which is done to ensure that only required returns enter the chain of recovery operations so that efforts and time do not get wasted. Further, collection involves gathering returns from different locations and transporting them to concerned warehouses for sorting. Sorting is done to separate the products having residual value from scrap which needs to be disposed off. Finally, relevant products re-enter the business cycle for value reclamation and are reprocessed through various recovery options e.g. remanufacture, recycle etc.

The efficient implementation of above mentioned phases increases the chances of successful returns management. However, success of RL also relies on customers as while organizations reclaim value of product returns, customers are required to deliver the returns to collection points so that they can enter the RL system. In addition to the mentioned environmental assistance, RL also provides following direct and indirect benefits: creates green image of the organizations which provides competitive advantage, reduces the investment in virgin raw materials which leads to increased revenue, facilitate proper handling and disposal of returns, helps in complying legislative regulations regarding product returns. (Curtis and Davis; Guide and van Wassenhove, 2001; Guo, 2012; Rogers & Tibben-Lembke, 2001; Choudhary and Madaan, 2013).

Beside the benefits, product recovery is a multifaceted and uncertain process as it involves unpredictable amounts and timings of the returns. Therefore, to deal with this complicated process of RL effectively, core competency and experience is necessary. As a result, most of the enterprises are going for outsourcing of product recovery operations and are opting for Third Party Recovery Facilitators (3PRF). Further, in order to get prominent results, selection of suitable 3PRF is essential. In this line, paper proposes a decision model to address the issue of suitable 3PRF selection. In previous researches, various techniques such as AHP, TOPSIS, and ANP etc. have been applied for selection of RL providers (Govindan and Murugesan, 2011; Meade and Sarkis 2002; Efendigil et al. 2008). The present research introduces a novel approach by combining ISM and ELECTRE III methodologies to determine the appropriate 3PRF. ELECTRE III is a more reliable method to handle this issue of 3PRF selection, as it ensures that an extremely low score on a criterion cannot be compensated by the good scores on other criteria (Wang and Lee, 2007). Unlike other methods, it is also a more practical and realistic technique as it also takes account imprecision present in the information. The study also considers a comprehensive list of criteria for selection procedure, some of which have not received the due consideration in available literature. The ISM methodology determines relationships among the various criteria to identify the most significant ones. Further, based on the most significant criteria, appropriate 3PRF is determined using ELECTRE III method. The application of this innovative approach can reduce complexity and time taken during the process of recovery facilitator selection. Additionally, case studies have been conducted to show the implementation of ELECTRE III methodology for 3PRF selection. The rest of the paper is organized as follows: Section 2 contains literature review, Section 3 explains ISM methodology, Section 4 deals with MICMAC analysis, Section 5 shows the implementation of ELECTRE III technique, Section 6 provides findings of the case studies and Section 7 contains conclusions and managerial implications.

2. LITERATURE REVIEW

Reverse Logistics is a value recovering process in which value of end-of life, used, defected, and obsolete products is reclaimed with the help of various product recovery operations (Guo, 2012). Initially, RL was mainly considered as a method of recapturing the value of product returns through recycling only. Nevertheless, scope of RL has evolved with time as amount of product returns has increased up to 50% of sales (Prahinski and Kocabasoglu, 2006) and presently, it is being implemented in different sectors of industries. Additionally, efficient management of product recovery operations results in enhanced customer relationships, conservation of virgin resources and an estimated reduction of about 10% in total annual logistics costs (Andel, 1997; Giuntini & Andel, 1995; Minahan, 1998). Despite these benefits, RL and its strategic value is often ignored by the organizations (Autry et al, 2001). The reasons for this reluctance towards RL implementation are: preoccupation with core business (Saccomano, 1997), Lack of awareness (Daugherty et al., 2001), intricacy in selection of suitable disposition option for handling the returns (Richey et al., 2005), difficulties in collection and transportation of returns (Baumgartner, 2000) and costs involved (Jayant et.al, 2012). Further, Meade and Sarkis (2002) stated that management of product returns can be difficult when arriving rate of returns is greater than recovery decision making rate as it can result in piling up of returns inventory. However, in present scenario organizations are obligated to take accountability of the returns due to extended producer responsibility and legislative regulations imposed by government (Aitken and Harrison, 2013). Consequently, due to these complexities and obligations involved, organizations are opting for outsourcing of RL and are taking the assistance of 3PRF.

Meade and Sarkis (2002) introduced the concept of third party recovery facilitators and implemented the Analytical network process (ANP) for selecting the suitable one. Accordingly, Guo (2011) proposed Analytical hierarchy process (AHP) for determining the appropriate 3PRF based on various criteria. Subsequently, Efendigil et al. (2008) and Govindan and Murugesan (2011) suggested the Fuzzy AHP and Fuzzy extent analysis for the same. Moreover, various techniques have been also proposed in literature for supplier selection in conventional supply chain (Kongar et. al, 2008; Li et.al, 2011; Lee et.al, 2012 etc.). Most of these Multi criteria decision making (MCDM) techniques present in literature are deficient in handling with the uncertainty and imprecision existing in the expert opinions. Additionally, some of the suggested decision models undergo information loss while processing, resulting in erroneous decisions (Liu and Wu, 2013). This paper suggests a novel MCDM technique, ELECTRE III, for selecting the suitable 3PRF as unlike other methods it considers the imprecision present in data and also ensures that an extremely low score on a criterion cannot be compensated by the good scores on other criteria. ELECTRE III is further integrated with Interpretive Structural Modeling (ISM) for identifying the most significant criteria in order to select the 3PRF in a more focused and reliable way.

The selection of suitable 3PRF is an intricate process as it involves comparison of various RL providers on the basis of different criteria. In order to assess the recovery facilitators, it is required to identify the decision criteria that would reflect multi-functionality of a 3PRF (Zhang and Feng, 2007). Moreover, there is a possibility of overlap between attributes for supplier selection and 3PRF selection. Therefore, quite a number of papers including literature on supplier selection in forward supply chain have been explored to identify the criteria which need to be considered during selection process. Table1 provides a summary of some of the papers studied.

Table 1: Summary of criteria considered in papers

S. NO.	AUTHORS	CRITERIA	SUB-CRITERIA
1)	Govindan et.al (2012)	Third Party Logistics Service, Reverse Logistic Function, Organizational Role, User Satisfaction, Impact of use of Third Party Logistic, Organizational Performance Criteria, IT Applications	Inventory Replenishment, Collection, Packing, Storage, Sorting, Reclaim, Recycle, Effective Communication, Service Improvement, Cost Saving, Quality, Cost ,Time ,Warehouse Management, Order Management, Supply chain planning etc.
2)	Chiou et. al (2012)	Environnemental Needs, Social Needs and Economic Needs	Recycled volumes, Recycling costs, Total manufacturing costs, Increase of sales volume for new products, Environmental regulations & directives, Consumers environmental awareness, Pressures with stakeholders, Reverse logistics management information system, Corporate social responsibility, Competitive pressures, Advertising promotion of image, Good recycling management system and recycling service
3)	Kongar et. al (2008)	On time delivery ratio, Confirmed fill rate, Service quality level, Unit operation cost, Capacity usage ratio, Total order cycle time, System flexibility index, Integration level index, Increment in market share, Research and development ratio, Environmental expenditures, Customer satisfaction index	
4)	Wang and Jhu (2012)	Enterprise Strength, Service Levels, Prices And Costs, Reverse Logistics Capabilities And Low-Carbon	Rapid Response Capability, On-time Delivery Rate, Network Coverage, Customer Satisfaction, Price, Cost Vehicle Condition, Peer Experience, Recycling Effect, Reuse Situation, Lower returns and Recall loss, Maintain customer Loyalty, Oil Consumption, Cleaning materials and clean energy use, Carbon Emissions
5)	Li et.al (2011)	Management Service, Technology Level, Information Level, Social Benefits and Ecological Benefits	Cost control ability, Communication and coordination ability, Recovery processing speed, Customer satisfaction, Industry experience, Technology innovation ability, Equipment leading level, Resource Utilization, Recycle level etc.

6)	Hong and Qiang (2009)	Service Level, Price, Enterprise Capacity, Experience And Enterprise Alliance	Reliability, Timely, Service Quality, Satisfaction degree of consumer, Financial investment capacity, Information technology capacity, Enterprise management capacity, Transport capacity, Industry service time, Supplying service class, Sharing of benefits and risks, Enterprise culture compatibility, Strategic concept compatibility, Management system compatibility
7)	Zhang and Feng (2007)	Strength, Union, Services, Experience And Price	Financial Ability To Invest, Transport Capacity, Network Coverage, Core Competencies, Strategic Concept Of Compatibility, Strategic Concept Of Compatibility, Service Quality System, Service Industries Time etc.
8)	Guo (2012)	Technology, Price, Speed And Quality	
9)	Cao and Zhang (2012)	Market Competitiveness, Service Ability, Information Level, Coordination Ability, Safety And Compatibility	Market share rate, Technical innovation ability, Service coverage area, Transportation capacity, Response to service requirements, Communication level with client, Degree of information monitoring, Financial stability, Compatibility of customer culture etc.
10)	Abdulrahman et.al (2014)	Management, Financial, Policy And Infrastructure	
11)	Meade and Sarkis (2002)	RL Functions, Organizational Performance Criteria And Organizational Role Of RL, Product Life Cycle Position	
12)	Lee et.al (2012)	Cost, Delivery, Flexibility, Quality, Technology And Risk	Pro-active in cost reduction, Accuracy of quantity fulfillment, Delivery condition, Perfect rate References from current customer, Reliability of quality, Optimization capabilities, Physical equipment, Financial stability etc.
13)	Peng (2012)	Cost, Operating Efficiency, Service Quality And Technology Level	Information technology, Storage technology, Transportation technology, Customer satisfaction , Culture compatibility , Enterprise credit, Operation speed, Operational readiness, Operation accuracy, Order processing cost , Storage cost ,Transportation cost
14)	Gol and Catay (2007)	General Company Considerations, Capabilities, Quality, Client Relationship, Labor Relations	Financial considerations, Location, Supply chain vision, Creative Management, Responsiveness Service, Quality and performance, Cultural fit, general reputation etc.

Although, a large number of attributes have been considered in literature for outsourcing of reverse logistics, still some factors have not received adequate attention till now, which include proper location of warehouses, optimum utilization of transportation, creative management, compatibility of information system and facility location site etc. These factors are essential to be considered in the selection process as they reflect the multi-functionality of 3PRF and its ability to perform the RL operations effectively and efficiently. Subsequently, on the basis of literature and discussions with the experts, a total of 10 criteria and 43 sub sub-criteria have been identified for selection of proper 3PRF. The experts are from the organizations implementing RL process in automobile and manufacturing sectors and as well as from academia. Table 2 gives a comprehensive list of criteria and sub- criteria considered in this work.

Table 2: Criteria and Sub-Criteria

S.No.	Criteria	Sub-Criteria	References
1)	Financial Viability (FV)	Financial Stability, Funds for Training, Transportation cost, Inventory carrying cost, Processing and Disposition cost	Kongar et. al (2008), Wang and Jhu (2012)
2)	Infrastructure (INF)	Sufficient in-house Facilities, RL Information System, Efficient Technology, Trained Manpower, Proper location, Geographical Spread, Range of Services Provided.	Göl And Çatay (2007), Li et.al (2011)
3)	Reverse Logistics Functions(RLF)	Screening, Collecting, Sorting, Reprocessing, Disposal	Govindan and Murugesan (2011), Meade and Sarkis (2002)
4)	Environmental Aspects (EA)	Proper Knowledge of Legislative Regulations, Creating awareness about Reverse Logistics, Optimum utilization of transportation, Proper Waste Management, Proper Utilization of Resources	Chiou et.al (2012), Abdulrahman et.al (2014)
5)	Social Aspects (SA)	Advertising green image of organization, Increase in market share, Proper Safety arrangement of workers, General Reputation, Experience in Similar Field	Cao and Zhang (2012)
6)	Service Level (SL)	Information sharing, Service Quality, Reliability, Coordination with the Organization	Peng (2012)
7)	Capabilities (CAP)	Risk management capability, Technology innovation capability, Supply chain vision, Creative management	Lee et.al(2012)
8)	Compatibility (COM)	Cultural compatibility, Strategic goal compatibility, Compatibility of information system, Linguistic compatibility	Hong and Qiang (2009), Zhang and Feng(2007)
9)	Efficiency(EF)	Cost reduction efficiency, Recycle and reuse efficiency	Govindan and Murugesan(2011)
10)	Effectiveness (EFC)	Facility location site, Recovery processing speed	Guo(2012)

To deal with entire maze of attributes, Interpretive Structural Modeling (ISM) is used for determining the relationships among criteria to identify the most significant ones. ISM is very helpful in such circumstances and provides a systematic method of arranging the attributes in hierarchal order.

3. INTERPRETIVE STRUCTURAL MODELING

Interpretive Structural Modeling (ISM) is a breakthrough for studying intricacies of a system and to understand various interactions among the components of the system. ISM portrays the interrelationship among the elements through a structural model consisting of links that define the direction of contextual relationship among the elements (Sohani and Sohani, 2011). Accordingly, ISM models assist in the transformation of vague, inadequately articulated mental models of systems into perceptible, precise models helpful for many purposes (Ahuja et.al, 2009). It has been widely used in several researches for determining the relationships among the variables (Saxena et al., 1992; Ravi and Shankar, 2005; Mathiyazhagan et al. 2013). Moreover, it is an appropriate modeling methodology to obtain the significance of various variables through evaluating the effect of one variable on other variables (Gupta et.al, 2012). The flow diagram of ISM is as depicted in figure 2.

Figure 2: Flow Diagram of ISM

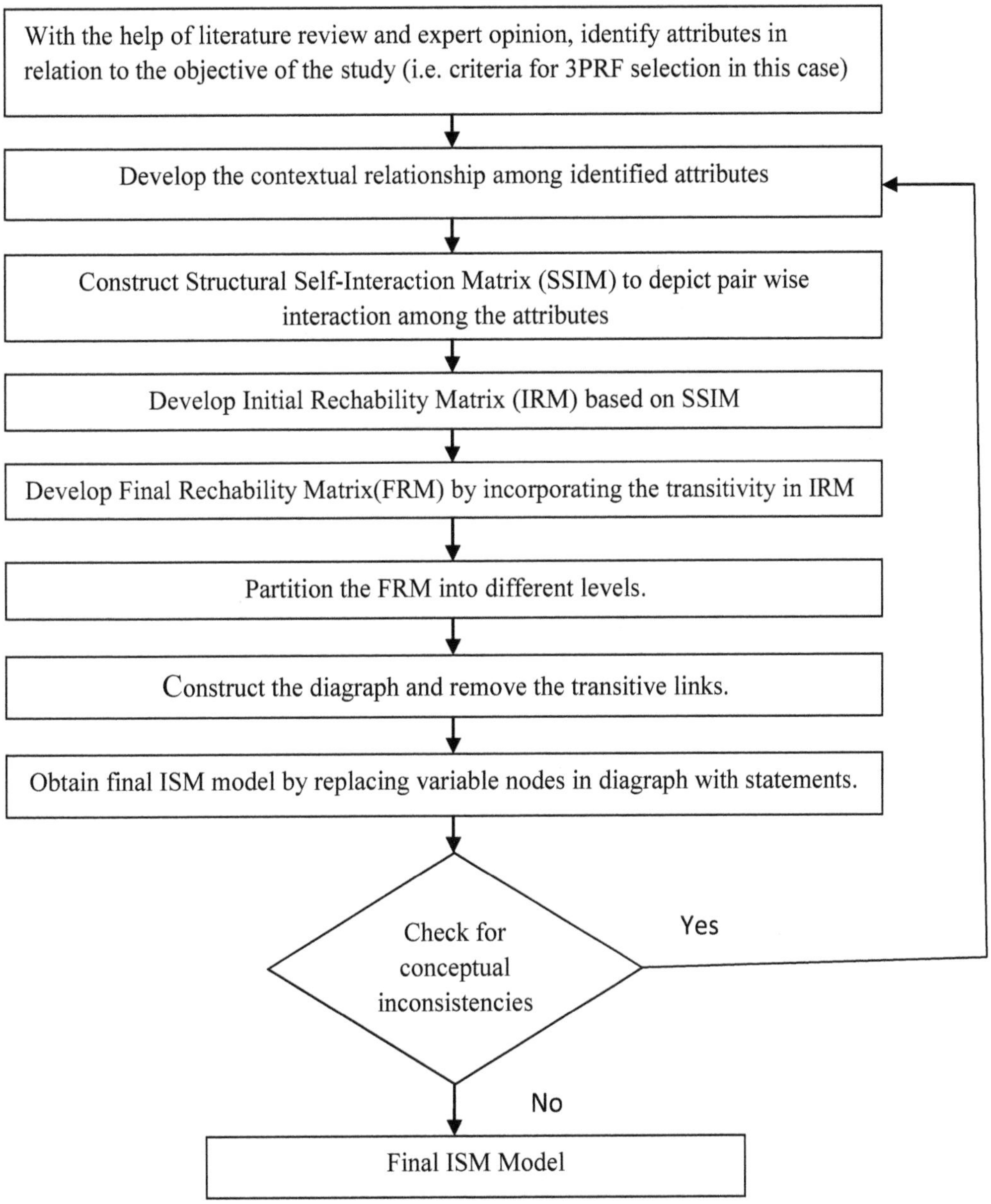

3.1 Application of ISM to determine the relationships among the identified criteria

After identification of 10 criteria and 43 sub-criteria through literature review, brain storming and expert opinion, analysis was carried out to determine the interrelationships among the various criteria. Accordingly, on the basis of steps explained in flow diagram, ISM is applied to find the interactions among the criteria given in table 2 for recovery facilitator selection.

3.1.1 Develop Structural Self Interaction Matrix (SSIM):

Opinions of 5 experts, 4 from automobile and manufacturing sector and one from academia, are used to develop the contextual relationships between different criteria by performing the pair wise comparisons. The symbols used to represent the relationship between the attributes are given below (i and j):

V – attribute i will help in achieving attribute j;

A – attribute i will be achieved by attribute j ;

X- attribute i and j will help to achieve to each other;

O - attribute i and j are unrelated

The SSIM constructed for 3PRF selection criteria using above described symbols is shown in Table 3.

3.1.2 Construct Initial Rechability Matrix :

Initial reachability matrix is developed from the SSIM by converting each cell entry into binary numbers "0" and "1". The substitutions are performed according to following rules:

- If the (i, j) entry in the SSIM is V, then the (i, j) entry in the reachability matrix becomes 1 and the

(j, i) entry becomes0.

- If the (i, j) entry in the SSIM is A, then the (i, j) entry in the reachability matrix becomes 0 and the (j, i) entry becomes1.

- If the (i, j) entry in the SSIM is X, then the (i, j) entry in the reachability matrix becomes 1 and the (j, i) entry also becomes 1.

- If the (i, j) entry in the SSIM is 0, then the (i, j) entry in the reachability matrix becomes 0 and the (j, i) entry also becomes 0.

The initial rechability matrix constructed for the criteria is shown in table 4

Table 3: Structural Self-Interaction Matrix (SSIM)

Factors	EFC	EF	COM	CAP	SL	SA	EA	RLF	INF	FV
FV	A	A	O	O	O	V	A	A	X	
INF	V	V	V	V	V	V	V	V		
RLF	V	V	A	A	V	V	V			
EA	A	A	A	A	O	V				
SA	A	A	O	A	A					
SL	A	A	A	A						
CAP	V	V	O							
COM	V	V								
EF	X									
EFC										

Table 4: Initial Reachability Matrix

Factors	EFC	EF	COM	CAP	SL	SA	EA	RLF	INF	FV
FV	0	0	0	0	0	1	0	0	1	1
INF	1	1	1	1	1	1	1	1	1	1
RLF	1	1	0	0	1	1	1	1	0	1
EA	0	0	0	0	0	1	1	0	0	1
SA	0	0	0	0	0	1	0	0	0	0
SL	0	0	0	0	1	1	0	0	0	0
CAP	1	1	0	1	1	1	1	1	0	0
COM	1	1	1	0	1	0	1	1	0	0
EF	1	1	0	0	1	0	1	0	0	1
EFC	1	1	0	0	1	0	1	0	0	1

3.1.3 Construct Final Reachability Matrix:

The final reachability matrix is obtained by incorporating the transitivity in the initial reachability matrix on the basis of transitivity rule i.e. if X is related to Y and Y is related to Z, then X is also related to Z. The final reachability matrix obtained is shown in Table 5

along with the driving power and dependence power. The driving power of a variable is the total number of variables including itself which it may help to achieve (i.e. row sum). The dependence power is the total number of variables including itself which it may help in achieving it (i.e. column sum).

Table 5: Final Reachability Matrix

Factors	EFC	EF	COM	CAP	SL	SA	EA	RLF	INF	FV	Driving Power
FV	1*	1*	1*	1*	1*	1	1*	1*	1	1	10
INF	1	1	1	1	1	1	1	1	1	1	10
RLF	1	1	0	0	1	1	1	1	1*	1	8
EA	0	0	0	0	0	1	1	0	1*	1	4
SA	0	0	0	0	0	1	0	0	0	0	1
SL	0	0	0	0	1	1	0	0	0	0	2
CAP	1	1	0	1	1	1	1	1	0	1*	8
COM	1	1	1	0	1	1*	1	1	0	1*	8
EF	1	1	0	0	1	1*	1	0	1*	1	7
EFC	1	1	0	0	1	1*	1	0	1*	1	7
Dependence Power	7	7	3	3	8	10	8	5	6	8	

3.1.4 Level Partitions

The reachability set and antecedent set for each criterion is determined from the final reachability matrix. The reachability set consist of the criteria itself and the other criteria, which it may influence, whereas the antecedent set (A) consists of the criteria itself and other criteria, which may influence it. Next, intersection set, which is the intersection of

corresponding reachability and antecedent sets are constructed. The criteria having similar intersection and rechability set are assigned top level position in the ISM hierarchy and are removed from the rest of the criteria. The procedure is iteratively repeated for the remaining criteria till levels are assigned to all the respective criteria. The first iteration is as shown in table 6.

Table 6: Iteration 1

Criteria	Rechability Set	Antecedent Set	Intersection Set	Level
FV	EFC,EF,COM,-CAP, SL,SA,EA,R-LF,INF, FV	EFC,EF,COM,CAP, EA,RLF,INF,FV	EFC,EF,COM, CAP,EA,RLF,INF, FV	
INF	EFC,EF,COM,-CAP,SL, SA,EA,R-LF,INF,FV	EFC,EF, EA,RLF,INF,FV	EFC,EF, EA,RLF,INF,FV	

RLF	EFC,EF,SL,SA, EA,RLF,INF,FV	CAP, COM RLF,INF,FV	RLF, INF, FV	
EA	SA,EA,INF,FV	EFC,EF,COM,CAP, EA,RLF,INF,FV	EA,INF,FV	
SA	SA	EFC,EF,COM,-CAP,SL, SA,EA,RL-F,INF,FV	SA	I
SL	SA,SL	EFC,EF,COM,-CAP,SL, RLF,INF,FV	SL	
CAP	EFC,EF,CAP,SL, SA,EA,RLF,FV	FV,INF,CAP	CAP,FV	
COM	EFC,EF,COM,SL, SA,EA,RLF,FV	FV,INF,COM	COM.FV	
EF	EFC,EF,SL, SA,EA,INF,FV	EFC,EF,COM,-CAP,RLF, INF,FV	EFC,EF,INF,FV	
EFC	EFC,EF,SL, SA,EA,INF,FV	EFC,EF,COM,-CAP,RLF, INF,FV	EFC,EF,INF,FV	

As it can be observed from table 6, Social Aspect (SA) has been assigned level I i.e. the top most position in the ISM hierarchy, which implies that SA will not influence any other considered criteria. Subsequently, after removing SA the entire iteration is repeated till rest of the criteria attain their respective levels. In this case, a total of seven iterations have been performed giving seven levels of ISM hierarchy. The iterations are shown in Table 7.

Table 7: Iterations 2-7

Criteria	Rechability Set	Antecedent Set	Intersection set	Level
EA	EA,INF,FV	EFC,EF,COM,CAP, EA,RLF,INF,FV	EA,INF,FV	II
SL	SL	EFC,EF,COM,CAP,SL, RLF,INF,FV	SL	II
FV	EFC,EF,COM,-CAP,RLF,IN-F,FV	EFC,EF,COM,CAP,RLF,IN-F,FV	EFC,EF,COM,-CAP,RLF,INF,FV	III
EF	EFC,EF,INF,-FV	EFC,EF,COM,CAP,RLF, INF,FV	EFC,EF,INF,FV	III
EFC	EFC,EF,INF,-FV	EFC,EF,COM,CAP,RLF, INF,FV	EFC,EF,INF,FV	III
RLF	RLF,INF	CAP, COM RLF,INF	RLF, INF	IV
CAP	CAP	INF,CAP	CAP	V
COM	COM	INF,COM	COM	V
INF	COM,CAP,INF	INF	INF	VI

3.1.5 Develop final ISM model

On the basis of final reachability matrix and the levels obtained in level partitioning, diagraph representing hierarchal relationship among the considered criteria is constructed. The directional interactions among the criteria are depicted with the help of arrows. Further, on eliminating the transitive links and replacing the nodes with criteria statements, digraph is converted into final ISM model as shown in figure 3.

It can be observed from final ISM model shown in figure 3 that Social Aspect (SA) criteria has appeared at the top of hierarchy and Infrastructure (INF) criteria has acquired the bottom level of the hierarchy. Consequently, it can be implied that Infrastructure is a crucial aspect, which needs to be emphasized while selecting the 3PRF as it appears on the bottom level of ISM model. It influences the other criteria on the upper levels of ISM.

Figure 3: ISM Model of Criteria

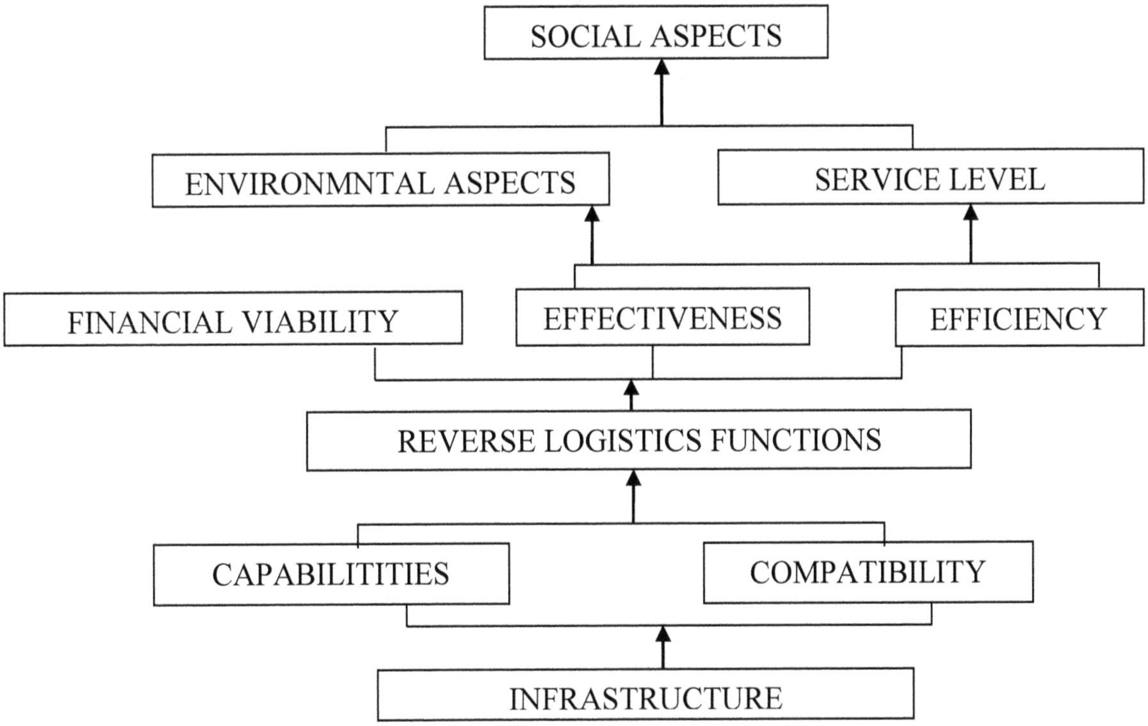

4. MICMAC ANALYSIS

MICMAC refers to Matriced'Impacts croises-multiplication applique' and classement (cross-impact matrix multiplication applied to classification) (Mathiyazhagan, 2013). The function of MICMAC analysis is to analyze and distinguish the variables on the basis of their dependence and driving power obtained from the ISM model. The driving power is the ability of a variable to influence or enhance the other variables, whereas dependence power is measure of the variable's tendency to get influenced by other variables. The driving power and dependence power are computed as explained in final reachability matrix section and are shown in table 5. On the

basis of MICMAC analysis, variables are divided into four categories as follows:

1. *Autonomous variables*: These are the variables which have weak driving power and weak dependence power. These variables are relatively disconnected from the system and have minimum influence. They are positioned in Quadrant I.

2. *Dependent variables*: This category represents those variables which have weak driving power but strong dependence power i.e. they will not drive other variables but will be strongly driven by others. They are represented in Quadrant II.

3. *Linkage variables*: The variables which have strong driving power and strong dependence power falls in this category. These are the unstable variables and any action on them will not only affect the other variables but may also have a strong feedback affect on them. They fall in Quadrant III.

4. *Independent Variables*: These are the variables which have weak dependence power but strong driving power. They are the key variables as they are not af-

fected by other variables but will play an important role in driving the other variables. They are placed in Quadrant IV

In this case, the driving power and dependence power of all the criteria considered for the selection of 3PRF are acquired from the final reachability matrix represented in table 5 and driving power Vs dependence power diagram is constructed as shown in figure 4.

Figure 4: Driving power and Dependence power diagram

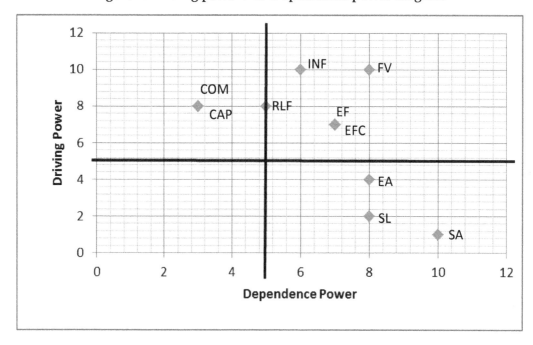

On the basis of ISM model and MICMAC analysis, it can be inferred that Infrastructure with driving power of 10 and dependence power of 6, is the most significant criteria that should be considered during the selection of recovery facilitator followed by Capability and Compatibility on the next level, then Reverse Logistics Functions on the third last level and so on. Further, it can be very tedious for organizations to consider all the attributes and sub-attributes during the entire selection process. So, the hierarchical model obtained with the help ISM methodology has been used for initial screening of criteria. Final selection of 3PRF can be done on the basis of attributes acquiring the bottom levels of model as they are the most significant ones. Moreover, it is believed that considering only important attributes in the final selection can provide a more focused view and would not have any adversarial affect on final results obtained. Thus, only the four most important

criteria namely, Infrastructure, Compatibility, Capability and Reverse Logistics Functions are being considered for the selection of 3PRF.

ELECTRE III method is being used for final selection of appropriate 3PRF based on the four most important criteria. In the literature, mostly methods such as ANP, AHP, DEA etc have been used singularly for the purpose of supplier selection. But, in this study ELECTRE III method together with ISM has been used to select the facilitator in a more precise and focused manner. ELECTRE III is a more reliable and realistic method for tackling such kind of problems, as it ensures that a very bad score on a criterion cannot be compensated by the good scores on other criteria (Wang and Lee, 2007) and decides the preferences on the basis of threshold values. A brief description of ELECTRE III is presented in the next section.

5. ELECTRE III

The acronym ELECTRE stands for: ELimination Et Choix Traduisant la Realité and is first proposed by Bernard Roy at SEMA consultancy company in 1965 (Roy, 1991). It is also applied in many other papers such as Sheppard et.al, 1999; Silva and almeida, 2012; Wang and Lee, 2007; Carlos, 2005 etc. ELECTRE is a multi-criteria decision analysis method that takes into account the uncertainty and imprecision, which are usually inherent in data produced by predictions and estimations. Traditional methods assume that the following two relations hold for two alternatives (a, b).

aPb (a is preferred to b) $g(a) > g(b)$

aIb (a is indifferent to b) $g(a) = g(b)$

(Where g(a) and g(b) are the performance of the supplier a and b respectively).

But sometimes there is a possibility that g(a) is only negligibly greater then g(b) and in that case we cannot say clearly that g(a) is preferable than g(b). For such kind of situations ELECTRE methods introduce the concept of an indifference threshold, q, and the preference threshold, p and give the following relations:

aPb (a is strongly preferred to b) $g(a)-g(b) > p$

aQb (a is weakly preferred to b) $q < g(a)- g(b) \leq p$

aIb (a is indifferent to b) $|g(a) - g(b)| \leq q$

(Where g(a) and g(b) are the performance of the supplier a and b respectively)

These two thresholds are defined as:

Indifference Threshold (q)–If difference between the performances of two alternatives is below indifference threshold, decision maker cannot make a choice between the two options.

Preference Threshold (p) – If difference between the performances of two alternatives is above preference threshold, decision maker can make a clear choice between the two options.

The ELECTRE methodology comprises two main procedures: the first part consists of construction of Outranking model and next step is to exploit the model and produce a ranking of suppliers from credibility matrix. Outranking relation is created with the help of following:

Concordance Index- It represents the extent to which the assertion aSb (a is at least as good as b) is valid for each pair of facilitators.

Discordance Index - It measures the degree of disagreement with the assertion aSb (a is at least as good as b) for each pair of facilitators with respect to each criteria.

For the calculation of discordance index another threshold called *Veto Threshold* (v) is required. Veto threshold can overrule the assertion aSb if there is any criteria for which 3PRF candidate B outperforms candidate A by at least the veto threshold i.e $gj(b)>gj(a)+vj$ where where gj(a) and gj(b) are the performance of the 3PRF A and B respectively for jth criteria and vj is the veto threshold for jth criteria.

5.1 Application of ELECTRE III Method for Supplier Selection

This section demonstrates the application of ELECTRE III method to select the most appropriate recovery facilitator. The data regarding the performance of the various 3PRF has been assumed hypothetically for the purpose of demonstrating the ELECTRE III methodology. The steps of ELECTRE III are as follows:

Step 1) Deciding the weights of all criteria which are under consideration for selection procedure.

A number of methods are present to calculate the weights of criteria. On the other hand, in some cases all the criteria are given same importance as decided by decision maker. In this case, Analytical Hierarchy Process (AHP) is being used to determine the weights of four most essential criteria, which are being considered for selection of 3PRF. A sample of concerned calculations is given in APPENDIX 1. The weights of the criteria thus obtained with the help of AHP and expert opinions are shown in Table 8.

Table 8: Weights of Criteria

Criteria	W
Compatibility	5%
Capability	10%
Reverse Logistics Functions	30%
Infrastructure	55%

Table 9: Decision table

Attributes	3PRF A	3PRF B	3PRF C	Weights
Compatibility	3	7	5	5%
Capability	5	7	3	10%
Reverse Logistics Functions	7	1	3	30%
Infrastructure	5	3	5	55%

Step 2) Constructing the decision table: For the evaluation of recovery facilitators with respect to the considered criteria, judgments are taken from experts handling the returns. The 3PRF are rated for each attribute on a scale of 1to 9 (1, 3, 5, 7, 9; very weak, weak, moderate, good, very good) and is shown in table 9.

Step 3) Deciding the *Threshold* values (Rogers and Bruen, 1998): Threshold values are subjective and variable. They depend on the task and situation under consideration. These values are decided by discussion with experts and are shown in table 10.

Table 10: Threshold values

Threshold	COM	CAP	RLF	INF
Indifference(q)	2	1	1	0.5
Preference(p)	4	3	2	1
Veto(v)	5	4	4	3

Table 11: Concordance Matrix

	3PRF A	3PRF B	3PRF C
3PRF A	1	0.9	1
3PRF B	0.15	1	0.15
3PRF C	0.65	0.90	1

Step 4) Calculation of *Concordance Index*: It is calculated with the help of following equations:

$$C\,(a,b) = \frac{1}{w}\sum_{j=1}^{m} w_j c_j\,(a,b) \quad -----\,(1)$$

Where A and B are two different recovery facilitator candidates, (j=1,.....,m) is the weight of each criteria, (j= 1,...,m) is the jth criteria, and

$$W = \sum_{j=1}^{m} w_j \quad ------\,(2)$$

The individual comparison indices for each criterion are calculated as follows:

$$c_j(a,b) = \begin{cases} 1; & g_j(a) + q_j \geq g_j(b) \\ 0; & g_j(a) + p_j \leq g_j(b) \\ \dfrac{p_j + g_j(a) - g_j(b)}{p_j - q_j}; & \text{otherwise} \end{cases} \quad ---\,(3)$$

Where and denote the indifference and preference thresholds, respectively, for criterion "j" and and are the assessment for the jth criterion of the candidates "A" and "B" , respectively. A sample of calculations made is given in APPENDIX 2. Similarly all the other concordance indexes are calculated and concordance matrix is given in table 11.

Step 5) Calculation of *Discordance Index*: It is calculated with the help of given equation:

$$d_j(a,b) = \begin{cases} 1; & g_j(a) + p_j \geq g_j(b) \\ 0; & g_j(a) + v_j \leq g_j(b) \\ \dfrac{g_j(a) - g_j(b) - p_j}{p_j - q_j}; & \text{otherwise} \end{cases} \quad ---\,(4)$$

Where U_j (j = 1......,m) is veto threshold. A sample of calculations made is given in APPENDIX 3.

Step 6) Credibility calculation: It measures the degree of outranking and is defined with help of following equation:

$$S\,(a,b) = \; C(a,b); \qquad d_j(a,b) \leq C(a,b)$$

$$= C(a,b). \prod_{j \in J\,(a,b)} \frac{1 - d_{j(a,b)}}{1 - C(a,b)}; \quad J(a,b)$$

Where J (a,b) is the set of criterion such that $d_{j(a,b)} > C\,(a,b)$

The credibility matrix, S obtained with the help of above equation is shown in table 12.

Table 12: Credibility Matrix

	3PRFA	3PRF B	3PRF C
3PRF A	1	0.9	1
3PRF B	0	1	0.089
3PRF C	0	0	1

Up to this is the construction of outranking model and the next step is to exploit the model and produce a ranking of 3PRF based on the outranking index. The general approach for exploitation is to construct two preorders Z1 and Z2 using a descending and ascending distillation process (respectively) and then combine these to produce a partial preorder Z = Z1 ∩ Z2.

Step 7) Calculation of matrix T defined as below:

$$T\,(a,b) = 1, \quad if \; S(a,b) > \lambda - s\,(\lambda)$$

$$= 0, \quad otherwise \qquad Where \quad \lambda = \max S(a,b)$$

A "credibility value" s( ), is determined such that only values of S that are sufficiently close to are considered. Thus if = 1, let s( ) = 0.15. For the given case the matrix obtained is as below:

$$T = \begin{matrix} 1 & 1 & 1 \\ 0 & 1 & 0 \\ 0 & 0 & 1 \end{matrix}$$

Step 8) Descending distillation:

Qualification for each facilitator is determined, which can be defined for 3PRF A, Q(A) as the number of 3PRFs that are outranked by 3PRF A minus the number of 3PRFs which outrank 3PRF A. In other words, Q is simply the row sum minus the col-

umn sum of the matrix T. The 3PRF having the highest qualification is the first distillate of D1. If D1 have only one 3PRF, previous process is repeated with A \ D1. Otherwise, the same process is applied inside D1. If distillate, D2 contains only one alternative, the procedure is started in D1 \ D2 (unless the set is empty); otherwise it is applied within D2, and so on until D1 is used up. The procedure is then repeated starting with A \ D1. The outcome is the first preorder Z1 that is the descending distillation.

For the first iteration the values of Q obtained are following:

Q(A) = 2 , Q(B) = -1 , Q(C) = -1

Since, Q(A) has the largest value so, 3PRF A will be the first distillate. Similarly, on performing the next iteration 3PRF B and 3PRF C will be the second distillate and will occupy the same ranking.

Step 9) Ascending distillation:

The ascending distillation is carried out in a similar fashion except that the facilitators with the smallest (rather than the largest) qualification are retained first.

In this case, the result of both the distillation is same so, the final ranking obtained is as below:

It can be deduced from the results that recovery facilitator A is the most suitable one as it occupies first position in ranking. Then, the facilitator B and facilitator C have same performance as they have obtained same ranking. But, still if we have to choose between the candidates B and C then it will depend on decision maker's preference or can be decided based on the performance of candidates in the most significant criterion.

6. CASE STUDY

Case study has been conducted in an automobile manufacturing organization with the intent to show practical application of suggested methodology for 3PRF selection and to discover significance of considered criteria in the process. The organization is planning to adopt the practice of RL to handle returns and is deciding upon outsourcing of recovery operations. Although, a large number of recovery facilitators are available, organizations are facing difficult to identify the most suitable one. However, organization has shortlisted three 3PRF candidates for final selection and has provided information regarding performance of each candidate in order to select the best one. The assessment of candidates with respect to each criterion (1, 3, 5, 7, 9; very weak, weak, moderate, good, very good) and weights of the criteria (taken from table 8) are given below in table 13. ELECTRE III methodology is applied to identify the most suitable 3PRF candidates as described in previous section.

Table 13: Data from the organization

Criteria	3PRF A	3PRF B	3PRF C	Wei-ghts
Compatibility	7	5	5	5%
Capability	9	5	7	10%
Reverse Logistics Functions	3	5	7	30%
Infrastructure	5	7	5	55%

Now, based on data given in table 13 and by applying steps of ELECTRE III method as described in section 5.1, ranking of 3PRF candidates is obtained and is shown below:

On the basis of ranking, it can be clearly deduced that recovery facilitator C is the most suitable one for the organization. Moreover, recovery facilitators A and B acquire same ranking position referring that both of them have almost same performance after 3PRF C. Hence, it can be seen that suggested methodology can practically tackle the issue of 3PRF selection without any ambiguity. Furthermore, the performance of all three 3PRF candidates is show below graphically in figure 5.

Figure 5: Performance of 3PRF Candidates

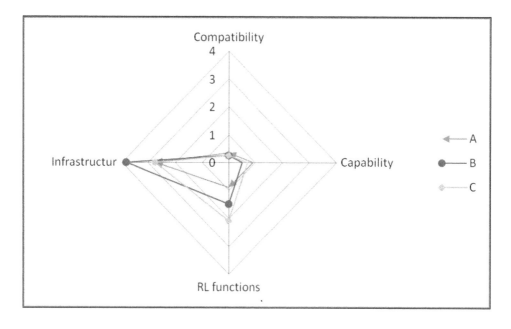

From the graph, it can be seen that 3PRF B has the best performance in area of infrastructure, which is the most significant criteria according to ISM model. Further, 3PRF A has the best performance related to capability but, very weak performance in the area of RL functions. On the other hand, 3PRF C, which is at first position in the ranking, has moderate performance in all the four considered criteria. So, it can be inferred that selection of 3PRF cannot be solely done on basis of performance in the most important criteria only. The performance in relatively less significant criteria also matters during the selection process.

7. CONCLUSIONS AND MANAGERIAL IMPLICATIONS

The product recovery operations are becoming imperative for organizations due to increasing legislative pressure and customer awareness. Additionally, due to complexity of RL activities, proficiency is required for its successful implementation. As a result, organizations are opting for outsourcing of recovery operations and therefore, issue of suitable 3PRF selection is arising. The paper introduces a novel approach combining the ISM and ELECTRE III methodologies to address this problem of 3PRF selection from a new perspective. ELECTRE III is a more focused and practical method, as unlike other methods it considers the imprecision present in data and also ensures that an extremely low score on a criterion cannot be compensated by the good scores on other criteria. The study presents a comprehensive list of ten criteria and forty three sub-criteria for initial screening of recovery facilitator candidates. Subsequently, the interrelationships among the criteria are studied with the help of ISM methodology to determine the most significant criteria. Additionally, the application of ELECTRE III method in the case study to determine the best 3PRF affirmed that suggested model has the capability to address this problem practically.

The proposed model attempts to focus and simplify the process of recovery facilitator selection by considering only the most significant attributes. In addition, the research also provides following implications at managerial level:

- The study successfully develops and implements a comprehensive decision model to guide organizations in the selection of optimal 3PRF.

- The model takes into account a wide range of criteria, making it more practical and realistic.

- In this study, none of the criteria is present in Quadrant I in MICMAC analysis which represents that all the considered criteria have a significant impact in 3PRF selection. Consequently, organizations should focus on all the considered criteria for initial screening of 3PRFs.

- Environmental aspect, Social aspect and Service level lie in the Quadrant II showing that they have very weak driving power and are dependent on other criteria, which implies that an enhanced performance in other considered attributes can influence the performance in these criteria also.

Accordingly, organizations can relax can a little while considering above mentioned three criteria.

- Capability and Compatibility are independent criteria as they have weak dependence power. Consequently, organizations need to emphasize more on these criteria as they are not getting influenced by other attributes.

- Reverse Logistics functions, Infrastructure, Efficiency, Effectiveness and Financial Viability are present in Quadrant III which makes them linkage variables having high dependence and driving power. These attributes are quite unstable as they highly influence other criteria and are also highly influenced by them. Further, organizations need to be careful while evaluating performance of 3PRF with respect to these criteria.

- On the basis of ISM model, Infrastructure, which includes sufficient in-house Facilities, RL Information System, Efficient Technology, Trained Manpower, Proper location and Geographical Spread, is the most significant criteria for 3PRF selection as it occupies the bottom most level of hierarchy.

- To reduce the time taken and tediousness of selection process, four most significant criteria acquiring the bottom levels of ISM model, Infrastructure; Capability; Compatibility and RL Functions, are suggested for final selection of recovery facilitator.

- The paper presents a novel MCDM technique, ELECTRE III, which is a more pragmatic and reliable technique for selection process as it takes into account imprecision and uncertainty which can exist in expert opinions.

- The proposed model can provide more focused and precise results compared to other models present in literature.

Although, organizations are moving towards recovery operations, still lot of work has to be done in this field to make RL a viable proposition. This paper addresses one of the issues associated with product recovery operations and can assist enterprises in taking one step closer towards RL. More case studies need to be carried out in different sectors e.g. manufacturing, electronics, automobiles etc. to get feedback from organizations for refinement and generalization of proposed model and results.

REFERENCES

Abdulrahman, M. D., Subramanian, N., & Gunasekaran, M. (2014). Critical barriers in implementing reverse logistics in the Chinese manufacturing sectors. *International Journal of Production Economics*, 147 B, 460–471

Ahuja, V., Yang, J., & Shankar, R. (2009). Benefits of collaborative ICT adoption for building project management. *Construction Innovation*, 9 (3), 323-340.

Aitken, J. & Harrison, A. (2013). Supply governance structures for reverse logistics systems. *International Journal of .Operations & Production Management*, 33 (6), 745 – 764.

Andel, T. (1997). Reverse Logistics: A Second Chance to Profit. *Transportation and Distribution*, 38 (7), 61-64.

Andel, T. & Giuntini, R. (1995). Reverse logistics role models. *Transportation & Distribution*, 36 (4), 97-100.

Autry, C. W., Daugherty, P. J., & Richey, R. G. (2001). The challenge of reverse logistics in catalog retailing. International Journal of Physical Distribution & Logistics Management, 31 (1), 26-37.

Baumgartner, J. (2000). Throwing it into reverse. *CED*, 26 (7), 101-105.

Bloemhof-Ruwaard, J., Krikke, H., & Van Wassenhove, L. N. (2004). OR models for eco-eco closed-loop supply chains. In Dekker, R., Inderfurth, K., van Wassenhove, L., and Fleischmann, M. (Eds.), *Quantitative Models for Closed-Loop Supply Chains*, Springer-Verlag, Berlin, Germany.

Cao & Zhang (2012). Research on the Evaluation of the Third Party Reverse Logistics Suppliers. *Advances in EECM*, 2 (140), 543–548.

Carlos, L. L. (2005). Multicriteria Decision Aid Application to a Student Selection Problem. *Pesquisa Operacional*, 25 (1), 45- 68.

Chiou, C. Y., Chen H. C., Cheng T. B. & Chun Y. Y. (2012). Consideration Factors of Reverse Logistics Implementation-A Case Study of Taiwan's Electronics Industry. *Procedia - Social and Behavioral Sciences*, 40, 375 – 381.

Choudhary, D. & Madaan, J. (2013). Hierarchical decision modeling approach for risk prioritization in sustainable supply chain. In the proceedings of International Conference on Humanitarian Logistics, IIM Raipur, India.

Curtis, G. & Jerry, D. Recovering Lost Profits by Improving Reverse Logistics. Available on http://www.ups.com/media/en/Reverse_Logistics_wp.pdf . Accessed on 20/12/2013.

Daugherty, P. J., Autry, C. W., & Ellinger, A. E. (2001). Reverse logistics: the relationship between resource commitment and program performance. *Journal of Business Logistics*, 22 (1), 107-110.

Efendigil, T., O'nu"t, S., and Kongar, E., 2008. "A holistic approach for selecting a third-party reverse logistics provider in the presence of vagueness." *Computers & Industrial Engineering* 54: 269–287.

Göl, H. & Çatay, B. (2007). Third-party logistics provider selection: insights from a Turkish automotive company. *Supply Chain Management: An International Journal*, 12 (6), 379 – 384.

Govindan, K. & Murugesan, P. (2011). Selection of third-party reverse logistics provider using fuzzy extent analysis. *Benchmarking: An International Journal*, 18 (1), 149 – 167.

Guide Jr., V. D. R. & Van Wassenhove, L. N. (2001). Managing product returns for remanufacturing. *Production and Operations Management*, 10, 142–155.

Guo, W. (2012). Selection Model of Third-party Reverse Logistics Service Providers Under supply chain management. In the proceedings of 24th Chinese Control and Decision Conference, Beijing, China, 1761-1764.

Gupta, P. M., Mangla, S. & Madaan, J. (2012). Multi-objective decision modeling using Interpretive Structural Modeling(ISM) for Green Supply Chains. In the proceedings of POMS 23rd Annual Conference, Chicago, Illinois, U.S.A., April 20- 23.

Hindustan Times (2013). Scrap Dealers to get you Good Deals or your Junks, April 11.

Jayant, A., Gupta, P. & Garg, B. (2012). Perspectives in reverse supply chain management(R-SCM): a state of the art literature review. *Jordan Journal of Mechanical and Industrial Engineering*, 6 (1), 87 – 102.

Kannan, G. & Haq, A. N. (2007). Analysis of interactions of criteria and sub-criteria for the selection of supplier in the built-order supply chain environment. *International Journal of Production and Research*, 45 (17), 3831–3852.

Kapetanopoulou, P. & Tagaras, G. (2011). Drivers and obstacles of product recovery activities in the Greek industry. *International Journal of Operations & Production Management*, 31 (2), 148 – 166.

Keong, L. M. (2008). Nokia kiosks collect phones for recycling. Available on http://news.cnet.com/8301-11128_3-10045417-54.html. Access on 13/01/2014.

Kongar, E., Efendigil, T. & Onut, S. (2008). A holistic approach for selecting a third party reverse logistics provider in the presence of vagueness. *Computers and Industrial Engineering*, 54 (6), 269–287.

Lee, C. K. M., William Ho, Ting He & Emrouznejad, A. (2012). Strategic logistics outsourcing: An integrated QFD and fuzzy AHP approach. *Expert Systems with Applications*, 39 (12), 10841–10850

Li, Zhou, J. & Wang, Y. (2011). Research on the Third Party Supplier of Reverse Logistics Selection under Low-carbon Economic Society. In the proceedings of International Conference on Mechatronic Science, Electric Engineering and Computer, Jilin, China, August 19-22.

Liu, H. C., L., and J. Wu. 2013. "Material Selection Using an Interval 2-tuple Linguistic VIKOR Method Considering Subjective and Objective Weights." *Materials and Design* 52: 158–167.

Meade, L. & Sarkis, J. (2002). A conceptual model for selecting and evaluating third-party reverse logistics providers. *Supply Chain Management: An International Journal*, 7 (5), 283 – 295.

Minahan, T. (1998). Manufacturers take aim at end of the supply chain. *Purchasing*, 124 (6), 111- 114.

Pinna, R. & Pier, P. C. (2012). Reverse Logistics and the Role of Fourth Party Logistics Provider. In Ales Groznik & Yu Xiong (Eds.), *Pathways to Supply Chain Excellence*, Intec, Italy.

Peng, J. (2012). Selection of Logistics Outsourcing Service Suppliers Based on AHP. *Energy Procedia*, 17A, 595 – 601.

Prahinski C. & Kocabasoglu C. (2006). Empirical Research Opportunities in Reverse Supply Chains. *Omega*, 34 (6), 519-532.

Richey, R. G., Stefan, E. G., & Patricia, J. D. (2005). The role of resource commitment and innovation in reverse logistics performance. *International Journal of Physical Distribution & Logistics Management*, 35 (4), 233 - 257.

Rogers, M. & Bruen, M. (1998). Choosing realistic values of indifference, preference and veto thresholds for use with environmental criteria within ELECTRE. *European Journal of Operational Research*, 107 (3), 542–551.

Rogers, D. S. & Tibben-Lembke, R. S. (1999). Going Backwards: Reverse Logistics Trends and Practices. University Of Nevada, Reno Center for Logistics Management, Reverse Logistics Executive Council, Pittsburgh, P.A.

Rogers, D. S. & Tibben-Lembke, R. S. (2001). An examination of reverse logistics practices. *Journal of Business Logistics*, 22 (2), 129-148.

Roy, B. (1991). The outranking approach and the foundations of electre methods. *Theory and Decision*, 31 (1):49-73.

Saccomano, A. (1997). Many happy returns. *Traffic World*, 249 (7), 22-24.

Sheppard, P., Vanderpooten, D. & Buchanan, J. (1999). Project Ranking Using Electre III. Northern Generation, Electricity Corporation of New Zealand, Université Paris Dauphine, France.

Silva & Almeida, M. (2012). Selection of rehabilitation construction solutions using ELECTRE III method University of Minho, School of Engineering, Department of Civil Engineering, Guimarães, Portugal.

Sohani, N & Sohani, N. (2011). Developing Interpretive Structural Model for Quality Framework in Higher Education: Indian Context. *Journal of Engineering, Science & Management Education*, 5 (2), 495–501.

Wang, J. & Zhu, Y. (2012). Research on Third-party Reverse Logistics Provider Selection Based on Fuzzy Clustering in Perspective of Low-carbon Economy. *CISME*, 2 (2), 63-66.

Wang, J. J. & Lee, F. H. (2007). Developing a Decision Model for Supplier Selection. In the proceedings of IEEE Conference, Shanghai, September 21-25.

Zhang, Y. & Feng, Y. (2007). A Selection Approach of Reverse Logistics Provider Based on Fuzzy AHP. In the proceedings of Fourth International Conference on Fuzzy Systems and Knowledge Discovery, Haikou, Hainan, China, August 24-27.

Zhi-Hong, Y. & Qiang, L. (2009). The grey comprehensive evaluation model of a third-party reverse logistics provider selection. In the proceedings of International Joint Conference on Artificial Intelligence, Hainan Island, China, 497-500

APPENDIX

APPENDIX 1

Calculation for determining weights of the criterion by AHP:

1. Pair wise comparison of the criteria

Criteria	Compatibi-lity	Capabi-lity	Reverse Logistics Func-tions	Infrastruc-ture
Compatibility	1	1/3	1/7	1/9
Capability	3	1	1/5	1/5
Reverse Logistics Func-tions	7	5	1	1/3
Infrastructure	9	5	3	1

$$A = \begin{matrix} 1 & 0.33 & 0.14 & 0.11 \\ 3 & 1 & 0.2 & 0.2 \\ 7 & 5 & 1 & 0.33 \\ 9 & 5 & 3 & 1 \end{matrix}$$

Normalized Column

$$\begin{matrix} 0.05 & 0.03 & 0.03 & 0.07 \\ 0.15 & 0.09 & 0.05 & 0.12 \\ 0.35 & 0.44 & 0.23 & 0.20 \\ 0.45 & 0.44 & 0.69 & 0.61 \end{matrix}$$

3) *Row Average* = X =
$$\begin{matrix} 0.045 & 0.05 \\ 0.1025 & 0.10 \\ 0.305 & 0.30 \\ 0.5475 & 0.55 \end{matrix}$$

4) *Calculation of consistency ratio*

$$AX = \begin{matrix} 0.19 \\ 0.42 \\ 1.33 \\ 2.4 \end{matrix} = X \quad so, = average\ of \begin{matrix} 3.8 \\ 4.2 \\ 4.43 \\ 4.36 \end{matrix} = 4.2$$

CI = (– n) / n-1 = 4.2 – 4 / 4-1 = **0.06 < 0.1**

So, the following results obtained are valid

Compatibility = 5%

Capability = 10 %

RL functions = 30%

Infrastructure = 55%

APPENDIX 2

Using the data from tables 8 and 9, calculation of concordance index is done with the help of equations 1-3 for candidates A and B as shown below: (Here w=1)

$c_1(A,B) = 0$; *since* $3 + 4 = 7$ $c_2(A,B) = 0.5$; *since* $(3 + 5 - 7)/(3 - 1) = 0.5$

$c_3(A,B) = 1$; *since* $7 + 1 \geq 1$ $c_4(A,B) = 1$; *since* $5 + 0.5 \geq 3$

$C(A,B) = (0 \times 0.05) + (0.5 \times 0.1) + (1 \times 0.3) + (1 \times 0.55) = 0.9$

APPENDIX 3

Considering the data in table 8 and veto threshold in table 9, discordance index for the criteria *Compatibility* for candidates A and B is calculated according to equation 4 as below:

$$d_1(A, B) = 0; \qquad since \ 3 + 4 \geq 7$$

Similarly, discordance index for all the other criteria for each pair of 3PRFs is calculated.

Incorporation of the Suggested Comments

1. Introduction - References are provided for the papers incorporating various techniques such as AHP, TOPSIS, and ANP etc. for selection of RL providers. (Pg – 4)

2. Literature Review

   -ELECTRE III technique is compared with other methodologies used in literature to justify its significance and advantages in 3PRF selection. (Pg – 6)

   -The importance of additional criteria considered for 3PRF selection is explained and information about the experts is provided. (Pg-8)

3. Method:

   - Various statements have been modified and even additional information is provided to better explain the Interpretive Structural Modeling(ISM). (Pg- 9-11)

   -The meaning of driving and dependence power is explained. (Pg-12)

   - The terms reachability set and antecedent set are explained and an additional table (Table 6: Iteration 1) with detailed description is provided to show how Level Partioning is done. (Pg- 13-14)

   - The importance of driving and dependence power in MICMAC analysis is specified. (Pg – 16)

   - Instead of two case studies only single case study is mentioned. (Pg-24)

4. Contributions: Some managerial implications are added and some are modified to better explain to show the implications of the findings. (Pg – 26-27)

5. Structure: The suggested comments have been incorporated in the structure and referencing of the paper.

AUTHOR'S BIOGRAPHY:

Divya Choudhary: Is currently working in the field of product recovery, reverse logistics, green supply chain management and risk management. She has completed her M Tech (specialization Product Design and Development) from MNNIT Allahabad. Presently, a research scholar in Department of Management studies, Indian Institute of Technology, Delhi, India.

Jitendra Madaan: Is currently working as Assistant Professor (Permanent) at IIT Delhi specialized in Operations Management. He has B.Tech, Mtech (NIT, Jaipur) and PhD in Industrial Systems from Indian Institute of Technology, New Delhi, He has covered a wide variety of areas in operations and information Systems Engineering and has a working knowledge to improve the information system performance. He has also been recently awarded with the DST young scientist grant. He has published several research papers on sustainable operations and its applications to various areas including role of IT in operations in national and international journals/conferences of repute including IJPR, POMS and DSS. He has guided one and two ongoing PhD's in the area exploring operational and decision framework.

Rakesh Narain: Has about 33 years of experience in teaching and research, industry and government department. His areas of research interest include Logistics and Supply Chain Management, Computer-Aided Manufacturing, Computer Integrated Manufacturing, Rapid Prototyping and Manufacturing, and Production and Operations Management. He got the *Literati Club 2004 Highly Commended Award* for contributing the best paper in Emerald journal in the 2003 volume. He has contributed research papers in refereed international journals and conferences, and has authored a chapter in a Textbook on Rapid prototyping.

9

Coordination in Brazilian Pharmaceutical Supply Chains

Priscila Laczynski de Souza Miguel
plsmiguel@gmail.com
FGV-EAESP

Manuel de Andrade e Silva Reis
manoel.reis@fgv.br
FGV-EAESP

ABSTRACT: Supply chain coordination (SCC) can be a challenge for many organizations as different firms in the same chain has different expectations and interdependencies (Arshinder & Deshmukh, 2008). Lack of SCC can result in the bullwhip effect and poor performance for a firm and its partners. By investigating the phenomenon in the Brazilian pharmaceutical supply chain using a qualitative research, this paper aims to understand the main issues that avoid a better integrated chain. Results of 21 interviews suggested that the lack of coordination in this environment was influenced by the network design and the history of the sector in Brazil, as well as scarce resources.

Key words: coordination, pharmaceutical supply chain, interorganizational relationships, bullwhip effect.

1. INTRODUCTION

Demand management (DM) is the Supply Chain Management (SCM) process responsible for balancing the productive capacity and the demand of a company, as well as the whole chain, synchronizing supply and demand between the upstream and downstream organizations. DM can contribute to firm performance, by reducing operational costs and inventories and increasing customer satisfaction and can only be effective if supply chain coordination (SCC) exists.

Supply chain Coordination (SCC) is a key construct in the Operations and Supply Chain Management field, although there is still no consensus on how it can be defined and its knowledge is still under development (Arshinder & Deshmukh, 2008). An effective management of activities and flow of goods requires that all individuals and organizations are aligned with supply chain goals and therefore, their actions should be coordinated (Fugate et al. (2006). According to Lee et al. (1997), the lack of coordination can result in the amplification of demand variability in a chain, also known as bullwhip effect that has been studied in the academic field since the seminal work of Forrester (1961).

Despite the successful cases, SCC is not a common practice, as managers face some challenges to extend coordination from within the organization to intraorganizational relationships: dominant member in the chain that takes decisions, different organizational culture, lack of coherent contracts, lack of meetings and commitment, incompatible information systems, conflicting objectives, independent cost evaluation of activities and processes, etc (Arshinder & Deshmukh, 2008). The former difficulties, however, disregard some important aspects of a supply chain that can result in obstacles for coordination, such as historical evolution of a supply chain and institutional aspects.

Therefore, this study aims to contribute to a deeper understanding of coordination barriers by investigating the reasons for the lack of coordination in the Brazilian pharmaceutical supply industry, that operates in the country since the beginning of 1900s and was affected by new regulations in the last 20 years as well as the entrance of new players.

Based on 21 qualitative interviews conducted with several organizations in the chain (suppliers, industry, distributors, drugstores, service providers and nodes representative associations), results suggested that lack of cooperation is a consequence of a chain that is highly influenced by the Brazilian institutional environment, low trust level between partners and lack of qualified professionals in the industry. The environment is highly regulated, resulting in few participants with high influence over the other echelons. Findings also provided evidence that the chain is not coordinated in an integrative way, but supply and demand are managed independently, with almost no information exchange.

The paper is structured as follows. Initially, the Literature Review section presents important references on supply chain management, demand management and coordination. Next, the methodology is presented and data analysis and key findings are discussed. The last section presents the main conclusions of the study, including its limitations and opportunities for future research.

2. LITERATURE REVIEW

2.1 The Supply Chain and Supply Chain Management

A supply chain is a sequence of activities, involving supply, manufacturing and distribution between many organizations, aiming to deliver value to a final consumer (Mentzer et al, 2001). Depending on the relationships between any dyad in a supply chain, companies adopt different governance mechanisms (from an arm's length to a more collaborative perspective) in order to maximize their benefits (Jarillo, 1988).

A more collaborative approach, also known as supply chain management (SCM), assumes partnerships between members of a chain, resulting in firm and overall performance improvements. The SCM assumes process integration, long-term relationships based on trust and commitment, collaboration in projects and processes and strategic information sharing between two or more organizations (Cooper et al, 1997; Cooper & Ellram, 1993; Mentzer et al., 2001). Although there is a discussion if SCM should be considered at the activity level (Lambert et al, 1998) or if it is a strategic and systemic approach (Mentzer et al, 2001), authors agree on the business process that should be considered: Customer Relationship Management, Customer Service Management, Demand Management, Order Fulfillment, Manufacturing Flow Management, Procurement, Product and Process Development and Reverse Logistics. The focus of the present research is on demand management process that is responsible for balancing demand and capacity in a supply chain.

2.2 Demand Management and the Bullwhip Effect

One of the key processes of supply chain management is demand management, responsible for balancing the demand and capacity of a firm and also the chain (Lambert and Cooper; Pagh, 1998). This process aims to balance the needs of customers with the operations of a firm and includes demand forecasting and synchronization between production, procurement and distribution, allowing the company to be more proactive for forecasted sales and react quickly and efficiently in case of unexpected demand (Croxton et al., 2002).

Taylor (2006, p. 164) defined demand management as:

> the study of the structure and efficacy of the information channels and procedures used to handle demand throughout the length of the supply chain, both internally within companies and externally between companies. It also includes detailed analysis of the nature of the demand information that is passed along these channels, including forecasts, orders, amendments and reconciliations.

DM has been studied since the seminal work of Forrester (1961) about industrial dynamics that stated that variations in production rates in companies are usually larger than the fluctuations in demand of final consumers. This difference would be caused by the amplification of small disturbances, as well as by the lack of information along the chain.

If processes within and among organizations are coordinated properly, the so-called "bullwhip effect" and therefore, the company's operational costs can be reduced (Metters, 1997; Lee et al., 1997). The bullwhip effect can be defined as the distortion of the demand as it is transmitted up the chain (Lee et al 1997, p. 93). As an example, imagine a chain with customer, retailer, wholesaler, manufacturing company, and a first tier supplier. Each echelon of the chain forecasts its demand, considering its own process, information and uncertainties and orders goods from the upstream side of the chain. The result is that the demand at the supplier is highly unstable, even if the customer demand is stable at the retailer. The consequences of the bullwhip effect are: high levels of inventory, low customer service level

and capacity and planning issues and constrains (Lee et al., 1997).

One of the main challenges for an efficient demand management is related to the interorganizational relationships. In order to share information, there must be trust and cooperation among its members, as well as coordination within an organization and between organizations. Otherwise, each firm will use safety stocks to protect against demand and supply uncertainties. (Lee et al., 1997).

Although several authors argue that the bullwhip effect cannot be avoided (Croxton et al., 2002; Dejonckhere et al .; 2003; Chatfield et al., 2004), there are several ways to mitigate it. Geary et al. (2006) list, based on extensive literature review, some principles that help to reduce the problem: control systems that measure the results of the chain, information sharing, reduced lead-times, shorter chains, synchronization between organizations, frequent demand forecasting and updating, joint planning with suppliers, optimization of batch production in order to reduce order quantities, avoidance of promotions and discounts ("everyday low price" strategy). Croxton et al. (2002) emphasize that, although the bullwhip is inevitable, the main goal of the demand management process is to implement some of the managerial practices that mitigate the issue and develop contingency plans to support a more effective and efficient customer service. To implement these practices along the whole chain it is important that all activities and processes are well coordinated in the SC.

2.3 Supply Chain Coordination (SCC)

Although SCC has been studied in the Operations and Supply Chain Management, it is still not a consensus between academics, with no unique definition (Arshinder & Deshmukh, 2008). According to Sahin and Robinson (2002, p. 3), "a supply chain is fully coordinated when all decisions are aligned to accomplish global system objectives..". In order to be effective, decisions in a supply chain should be rationally taken to optimize supply chain results (Chen & Xiao, 2009) and all activities and flow of goods and information have to be managed and controlled in order to create value to the final customer (Fugate et al, 2006). A literature review on the theme identified the definition of Malone and Crowston (1994) as the most commonly accepted: "the act of managing dependencies between entities and the joint effot of entities working together towards mutually defined goals" (Arshinder & Deshmukh, 2008, p. 318).

It is important to highlight that coordination does not mean cooperation. While the previous concept is related to interaction, exchanges and alignment between members of a supply chain, the former is related to mutuality and collaboration and assumes a more close and commited relationship (Santos & D'Antone, 2014). A coordination system defines processes and responsibilities inside the firm and at the interfaces of supply chain (Arshinder & De-shumkh, 2008).

There are some key points that integrate the field: a) SCC assumes that there are different interdependencies and different type of conflicts between SC members; b) it is hard and complex to manage processes and activities between these organizations; c) there are different mechanisms that can be adopted in different chains and relationships; and d) there is a need to control if the coordination mechanisms is really effective (Arshinder & Deshmukh, 2008).

Some of the mechanisms that can be used in a supply chain to implement coordination are supply chain contracts, information sharing, joint decision making and information technology (Arshinder & Deshmukh, 2008). Li and Wang (2007) complement saying that centralized and decentralized supply chains require different mechanisms of coordination. While a centralized supply chain, that takes a systemic view with the main goal of a superior overall performance, can implement an action plan to control and integrate all activities and decisions, this is not true in a decentralized supply chain, where members adopt a more opportunistic behavior. In the last case, an incentive schema is necessary to motivate firms to cooperate (Li & Wang, 2007). Xu and Beamon (2006) corroborate emphasizing that the appropriate coordination system (resource sharing structure, decision style, level of control and risk and reward sharing) depends on the interdependencies between organizations, uncertainty and information technology and results in different costs associated with coordination (coordination costs, operational risks costs and opportunistic risk cost).

The benefits of coordination are well known in the literature: inventory reduction, leadtimes reduction, increased customer satisfaction and revenue (Lee et al, 1997). However, managers fail to implement effective coordination due to different reasons that encompass human and procedures issues.

At the human perspective, coordination assumes interaction (i.e., good communication and personal relationships), exchange of information and resources and alignment to common goals (Santos & D'Antone, 2014). However, many conflicts exist in the chain, such as individual goals and objectives, disagreement over leadership and different perspectives on decision making. Companies also differ in terms of organizational culture. On the other hand, SCC also depends on tools to monitor the actions and processes, such a unique performance measurement and cost evaluation systems; coherent contracts; frequent and regular information sharing; compatible information systems and agreement on practices such as order quantities, replenishment systems, frequency of deliveries, production cycles and batch sizes (Arshinder & Deshmukh, 2008).

One gap identified in the SCC literature refers to how coordination can be affected by the evolution of the whole SC and external aspects such as institutional forces (Tang et al, 2008). According to Yaibuathet et al (2008, p. 262), "the IE (institutional environment) is one of the most significant factors for the achievement of SCM development" and considers both national settings and context. It is important to consider the relationship between institutions and the external environment, as organizations operate in different levels: country level, industry/sector level (macro), supply chain level (meso) and the firm level (micro) (Kinra & Kotzab, 2008). The IE encompasses demands of government and regulatory bodies (regulative elements), society values and norms, normally associated to a country or region culture (normative elements) and cultural-cognitive that refers to a previously established "framework of individual society" (Yaibuathet et al. 2008, p. 263). Previous researches have provided evidence that IE influences integration (Cai et al., 2010; Wong & Boon-itt, 2008), technology adoption (Zhang & Dhaliwal, 2009) and highly influences the SC performance (Kinra & Kotzab, 2008; Yaibuathet et al., 2008).

Bearing this in mind, this study aims to investigate the barriers for the lack of coordination in the Brazilian pharmaceutical supply chain, as discussed in the next section.

3. METHODOLOGY

A qualitative case study methodology was adopted as the research strategy in order to accomplish with the objectives. A case study can be defined as a as an empirical research methodology that investigates a contemporary phenomenon within its real context, especially when boundaries between phenomenon and context are not clearly evident (Yin, 1984). A deductive approach was selected with the purpose of theory elaboration (Kekovivi & Choi, 2014), as the idea was to extend the previous knowledge about the phenomenon (coordination) complementing it with an institutional environment perspective (Barratt et al., 2011).

The unit of analysis in the present research was the supply chain as a whole. The Brazilian pharmaceutical supply chain (BPSC) was elected as the case for two main reasons. First of all, previous research has provided evidence of bullwhip effect and lack of coordination in Brazilian pharmaceutical chain (Miguel et al., 2009). As we would like to explore the hindrances of SCC, this is a obligatory requirement. Second, BPSC is one of the most important industries in Brazil and operates in the country since the 1900. Despite this, it is a very dynamic environment, with the entrance of new players and the influence of the government. The case is described in the next topic.

3.1 Case description: Brazilian Pharmaceutical Supply Chain

The pharmaceutical industry is one the most profitable sector in the worldwide economy, especially due to its high innovation rates, that represent a barrier to new entrants, either due to the process of patents registration or due to the high investments in research and development (R&D) required. It is also a regulated industry with high quality and reliability standards. Due to the limited number of players, the barriers to new competitors and the high profit margin, the industry does not prioritize efficiency (Capanema, 2006).

In Brazil, the pharmaceutical industry is responsible for an income of U$ 25,5 billion in 2012 (ABIQUIM, 2013). It is a highly fragmented sector, but with sales concentration in a few players, mainly multinational companies - 60% percent international companies and 40% local companies according to Magalhães et al (2008). The operations consist normally in the production and distribution of drugs, with raw material being imported from other countries. Innovation activities are typically carried out abroad (Bastos, 2005).

The industry is highly regulated and prices are controlled by the government since 2001 (Romano et al, 2007). Products are also highly taxed - approximately 25% of sales revenue (Amaral, 2006).

In 2000, the Brazilian government authorized the production and sales of the so called generic drugs (i.e. a similar product or drug, which the patent protection has expired), increasing the competition of the sector and eliminating one of the most important barriers to new entrants (Capanema, 2006). In order to be competitive, leading companies have developed their own generic operations.

Figure 1 presents the Brazilian pharmaceutical supply chain. Industry is the focal company that is responsible for developing, producing and distributing the products in the market. The main suppliers are the chemical and pharma-chemical companies, as well as packaging suppliers. In Brazil, as the focal companies are normally subsidiaries of multinational companies, they also import raw material from other operations of their own corporation. Finished goods are distributed to retailers, through distributors and drugstore chains and are also sold to public and private cooperatives.

Figure 1: Brazilian Pharmaceutical Supply Chain

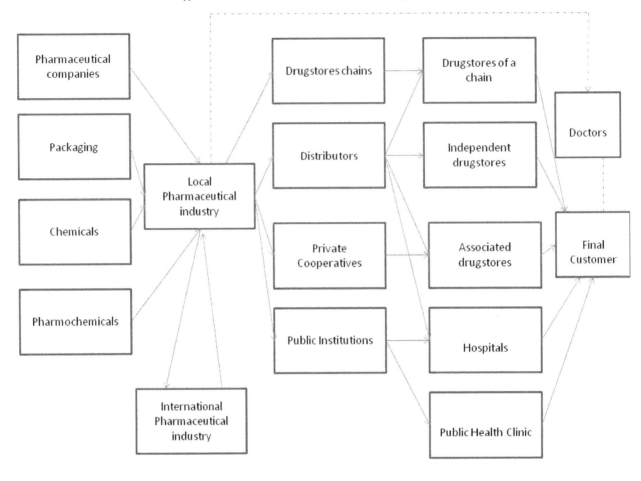

Source: Adapted from Miguel et al. (2009)

The pharmaceutical supply chain is structured in basically four nodes before the final consumer: Suppliers, Focal Company, Distributors and Retailers. In Brazil, there are between 50.000 to 70.000 point of sales (POS), highly sprayed in the whole country.

Although the pharmaceutical chains have national coverage, each geographic region presents different characteristics. The manufacturing operations are located basically in the Southeast region (Especially São Paulo and Rio de Janeiro). Major distributors have their operations in the Southeast, Midwest and Northeast regions, while in the North region, local distributors are responsible to fulfill orders to the point of sales, due to the logistical and transport constrains. From their Distribution Centers, the distributors are responsible to ship products to small and independent retailers.

With increased participation in the total market share, the drugstore chains are located in large cities

and centers (mainly in SP and RJ). They purchase a large amount of goods from the focal companies and replenish inventory of their POS.

3.2 Data collection and analysis

During the research, 21 qualitative interviews with various companies and organizations in the supply chain were conducted from April 2010 to February 2011 (See tables 1 and 2 for the profile of respondents). Besides the main players in the supply chain such as suppliers, focal companies, distributors and drugstore chains, interviews were also conducted with 3 representative associations to have a broader view of the chain and with 2 service providers that assisted the companies with the Sales and Operations Planning procedures. Interviews took an average of 90 minutes and were recorded upon agreement of the respondents and transcribed for data analysis.

Table 1: Profile of organizations

Echelons in the supply chain	Number of interviews
Suppliers	4
Focal companies	4
Distributors	4
Drugstore chains	4
Representative associations	3
Service Providers	2
Total	21

Source: Authors

Table 2: Profile of respondents

Department	Number of interviews
Business Strategy	4
Procurement	5
Sales/Commercial	3
SCM/Logistics	5
Operations	1
Finance	1
Consultant	2
Total	21

Source: Authors

During the data collection process, a semi-structured interview protocol was used (Appendix I). The interview consisted of two sections: 1) Questions about internal and intra-organizational relationships; 2) Questions about the internal process of each organization. No direct question on coordination was included to avoid bias from the respondents.

Each interview was analyzed concomitantly with the data collection in three steps. First, main ideas were extracted from the interview transcription; sec-ond, a more detailed analysis of the data identified context, condition and agents interactions and previous codes (A list of codes is presented in appendix II) were assigned ; finally, previously defined and new codes were classified in terms of the power of evidence provided (weak, moderate and strong). After each interview was analyzed, data were crossed to verify patterns, similarities and divergences (Collis & Hussey, 2005, p. 247)

4. DATA RESULTS

The results of the data analysis were reported in three sections: Historical Evolution of the Brazilian Pharmaceutical Supply Chain, Internal Process and Intra-organizational relationships.

4.1 Historical Evolution of the Brazilian Pharmaceutical Supply Chain

Although the supply chain in Brazil is not new, it is in constant development. According to the interviewees, until the 80's, the focal companies were responsible for distributing products directly to the point of sales. Due to the increase of retailers and the need to invest in R&D, there was a need to reconfigure the chain and the distributor node was included. The distribution function was outsourced and many employees of the industries started to lead the new operations and act as service provider.

In order to have economies of scales, the distributors could not be exclusive and they had to start working with different focal organizations. At this time, Brazil was facing high inflation and distributors started to make profit from speculative inventories: they purchased large quantities from the manufacturing and resold to retailers at a higher price. An additional incentive to the distributors was the fiscal war between the different regions of the countries. Their strategies were to prioritize fiscal gains instead of logistics efficiency. That resulted in a complex and unbelievable situation that made goods travel huge distances, without any need, increasing leadtime and therefore inventories at the different stages of the chain. At this time, distributors had three main activities: logistics, sales and credit that was a particularity of the Brazilian environment.

Until the decade of 1990, the pharmaceutical retailers were small point-of-sales, basically familiar business, with low professionalism. Those POS were called independent drugstores. In the 2000, however, those independent businesses started to form cooperatives or to be part of a drugstore chain. The existence of the independent POS, besides the lack of control, also promoted a parallel market of goods, with no control of inventories. Recently, however, the Brazilian government implemented the Electronic Fiscal Invoice and a new fiscal regime ("tax substitution regime") that forced the players to improve their processes and also to control better their business. Therefore the organizations that were making profit with the lack of control lost their bargaining power and size and had to redefine their strategies.

The distribution node also presented another peculiarity: the high competition between the organizations and the lack of practices and professional management resulted in a "just in time" customer service to retailer. According to respondents of this research, distributors had to deliver goods to POS daily, without a good system to manage inventory and transportation, resulting in higher operational costs and reduction of profitability. Many players in the industry had broken in the past years.

The industry was also been challenged with the new competition of generic drugs introduction. The reference industry lost their bargaining power, because distributors and suppliers now had new alternatives in the market. In order to overcome the competition, the focal companies were implementing two new strategies: 1) They were increasing their volume of sales to the drugstore chain, without the intermediation of the distributors; 2) They were implementing a new strategy, called Logistics Operations, in which their own sales representatives visit the biggest independent drugstores and offer products directly. When the POS places the order to the focal company, the order fulfillment is done through the distributor that is responsible only for the logistics operations. Although distributors claimed that the operation was not fair, the industry argued that they would like to be closer to the final consumer.

According to one of the independent drugstores, the logistics operation presents a trade-off between higher transparency and logistics organization and their ability to bargain in the purchasing process. According to one respondent, the practice improved the process but resulted in lower profitability.

Local suppliers of chemicals raw-materials were rare. In fact, the industry was almost broken due to the lack of governmental incentives and international competition. Although new players were limited due to the ANVISA certification, products were imported from other countries, because import taxes were low, according to the interviewees. The packaging industry, on the other hand, was local and not exclusive. Packaging providers also supplied products to the Food and Cosmetics industries. According to the respondents, "the volume that they supply to the pharmaceutical industry is just a small portion of their portfolio, but it is highly profitable."

The dynamic evolution of the pharmaceutical supply chain and the regulation of the sector resulted in an environment that did not promote close relationships and efficiency, but foster opportunistic behavior. The new competition scenario with the entrance of new players (generic industry and drugstores chains) forced a new configuration of the network and also changed the bargaining power of the original players. In order to survive, they were competing against each other and were not open to collaboration. The pharmaceutical supply chain can be defined as a decentralized supply chain (Li & Wang, 2007).

4.1 Internal Process

In the studied chain, the majority of the organizations did manage supply and demand independently and did not have process to synchronize them, although in some cases, both supplier management and demand management processes were functions developed in the same department. The respondents' reports provided evidence that companies were mainly concerned to achieve internal goals and were not focused on the chain as a whole.

According to one interviewee, although the whole industry was getting aware about logistics and supply management, there were still major conflicts between different departments within the company, avoiding internal integration. The employees of the organizations did not have a systemic and strategic view and had no interest to understand and to learn about the processes and activities performed by a different function, what might result in inefficiency. Another respondent believed that there was resistance inside organizations to promote internal integration. As he mentioned "people (in the industry) are not prepared to a shift in their own paradigm." The industry looked for new initiatives that would promote integration, but did not prepare their employees. Some of the reasons mentioned in the interviews were: no alignment with the HR department, lack of top management support and a short term view.

Another aspect that was constantly mentioned during the interviews was the lack of expertise in the industry, particularly in terms of logistics and planning. Several respondents mentioned high turnover in their organizations, driven by the competition with the generic manufactures. To reinforce the point, many of the respondents were pretty new in the organizations. Especially in the distributors, it was possible to notice that many of them were facing an internal reorganization and were hiring new employees from different sectors, aiming to increase competitiveness and bargaining power.

The interviews suggested that the pharmaceutical chain is deprived of skilled labor in logistics processes and supply chain management practices, although many companies were working to implement different tools. Only with the development of internal skills, the industry might achieve higher forecast accuracy and minimize the bullwhip effect. Processes such as Sales and Operations Planning (S&OP), which were already being deployed in companies, would allow organizations to achieve internal alignment and consequently, external collaboration. Top management support would also be key in this process.

To minimize the bullwhip effect, the focal companies were taking two actions: a) contracting service providers that gathered information from retailers directly; b) improving their inventories control of finished goods, both internally and at the distributors, although inventory is still used as a bargaining tool. In terms of raw and packaging material, their main issue is related to imports clearance.

The distributors were also taking actions to survive. Driven by the challenges they had been facing, they were looking for new organizational structures and were reviewing processes to provide a new type of service.

Considering the discussion above, we can conclude that, because there is no internal integration in the companies, it is even harder to work with other organizations in the chain and therefore achieve SCC.

4.2 Internal and Intra-organizational relationships

As discussed before, the Brazilian pharmaceutical chain has not been planned for logistic efficiency or responsiveness and it was highly influenced by institutional aspects that provided higher bargaining power of the different links in the chain in different sectors.

The research suggested that the chain is characterized by lack of trust between the various links and almost none internal synchronization in business. In all interviews, it was found that there was no intention to cooperate with the other companies, although this was recognized as a best practice. The dispute for bargaining power between the various players and competition between firms within the same link are reflected in the lack of transparency and information sharing between the organizations.

The interviews also revealed that the government had influence on their operations, since regulation and tax regimes adopted in Brazil resulted in the need to take reactive actions. As examples, the law of generics increased competition and decreased the power of the reference laboratories and the increased fiscal control reduced the size of the distribution node.

As the interviews were conducted in the year of election to the presidency of the country, many respondents expressed uncertainty regarding measures that would affect the industry, such as the obligation to maintain the medicines behind the counters or the program called Popular Drugstores, created during the government of President Lula. According to several respondents, although the program has been presented as an alternative to the low-income population, the operationalization of the program has several limitations. One of the constraints was the certification of the authorized point of sales, as each outlet should be certified instead of the drugstore chain. According to the distributors' point of view, the government fostered closer relationships between industry and retail, disregarding the presence of distributors and encouraging the industry to operate directly with the retailer channel.

The conflict between industry and distributors was evidenced in the majority of the interviews with respondents of these ties. Although the industry was looking for logistics performance improvements using data collected directly with the retailers, they did not consider working closer to the distributors. Just as one example, one respondent mentioned that they were targeting to reduce days of inventory in the whole chain (industry and distributors), but they did not discussed that with their customers.

When they believed that the inventory is high at the distribution level, they just interrupted the supply to them, without discussing products with excess inventory or shortages. This kind of approach is a consequence of their past history. Even when they claimed that they were trying to implement a closer relationship with drugstore chains, they did not provided any evidence that they were looking for integration. Companies were more focused on achieving internal goals.

According to the service providers interviewed, the goods are pushed through the chain and the lack of logistic expertise and an agreed leadership resulted in such conflicts. Just recently the players were getting aware of some of the best management practices discussed in the literature.

On the suppliers side, the lack of collaboration is also evident, but for different reasons. According to some of the interviewees, the volume of raw and packaging material that is sourced locally is quite low, compared to the total cost of material. Therefore, the industry does not have interest on developing and maintaining close relationships. The relationships, however, are less conflicting, perhaps due to the lower mutual dependence between the organizations. The qualified suppliers are protected by the certification process in the pharmaceutical supply chain. As ANVISA (the local FDA organization) control all products in this industry, the dyad can be classified in a "lock-in" situation.

It can be concluded that the existence of several participants in the pharmaceutical chain, competing against each other, has resulted in an associative or asymmetric supply chain, with high redundancy, i.e. there were many organizations supplying the same products and services, therefore, achieving the same costs, with low switching cost between suppliers / customers (Crook & Combs, 2007; McCarter & Northscraft, 2007). In this type of chain, each organization does its own activities independently of the other members of the chain and the bargaining power is usually concentrated in a single and stronger company, which appropriates the most gains of the supply chain.

According to Crook and Combs (2007), value can be created in a supply chain by each organization that contributes to individually and an extra share of value can be created by the synergy and interaction of the firms. The total value can be appropriated by the strongest organization or can be shared between each organization depending on the bargaining power of each party (Brandenburger & Stuart, 1996, Coff, 1999; Bowman & Ambrosini, 2003). This bargaining power depends on the resources available and control in the chain, as well as its structure and number of organizations (concentration) (Crook & Combs, 2007). If the firm has access to information, it can estimate the interdependencies of all members and it might capture a larger share of the gains during negotiations (Crook & Combs, 2007). In the studied case, because the bargaining power is under discussion, the organizations were not willing to cooperate and coordinate activities.

According to the interviews, the most profitable tie in the chain was the focal companies, responsible for the product development and manufacturing. Their high margins were the value appropriated by them and it was created by the industry individually. Local suppliers did not add much value to the final product, as the most important raw material came from foreign countries. The distributors and drugstore chains margins were also quite irrelevant to the total value created, and their value creation came from logistics and fiscal processes.

As the focal company could only deploy its products to the market through distributors for a long time, the bargaining power of the distributors increased and the focal companies had to share the leadership and the gain with this tie. However, the status of the distributors weakened when the inflation got controlled, the government started to improve the fiscal control and the industry had a new alternative to sell product (drugstores chains). The focal company, also, was impacted by the release of generic production in Brazil. So both echelons in the pharmaceutical network had to readapt to a new scenario to recover their gains and were working individually, competing with the other tie to increase their own bargaining power.

The history of the chain had also shown that companies always adopted an opportunistic behavior, resulting in the lack of trust. Therefore, there was no information sharing, because access to information could mean more bargaining power and more value appropriation. The companies did not foresee the benefits of working in a collaborative manner.

The previous results can be regrouped according to the literature of SCC and institutional perspectives on SCM as shown in figure 2. Besides the traditional barriers, this research provided evidence of news force that inhibits the SCC: the dynamics of the environment that forces the organizations to adapt themselves to new constrains in the system in order to achieve a new optimal point and the influence of the macro infrastructures on the SCM, such as politics, economy culture, that is better understood by institutional perspectives (Tang et al. 2008).

Figure 2: Barriers to SCC

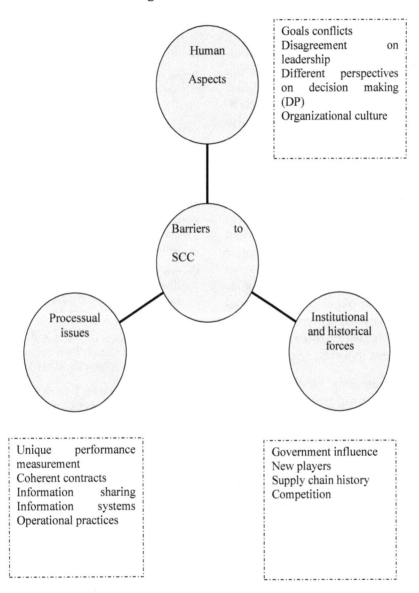

The proposed framework can also be analyzed according to the institution level as suggested by Kinra & Kotzab (2008). While the traditional barriers exist at the micro or meso level (i.e, organization and supply chain), the new forces refer to the macro level (country level) that influences the environmental complexity and change the constraints perceived by each firm (Kinra & Kotzab, 2008).

5. CONCLUSIONS

Aiming to understand the lack of coordination and the resulted bullwhip effect in the pharmaceutical supply chain, this research adopted a case study approach. The results suggest that the mainly causes of a poor demand management process were: a network designed to achieve fiscal efficiency instead of logistics efficiency, dispute for bargaining power in order to appropriate more value, resulting in low levels of confidence and lack of internal resources and skills to better manage processes internally and with other companies in the chain.

Compared to challenges to SCC mentioned in the literature section - dominant member in the chain that takes decisions, different organizational culture, lack of coherent contracts, lack of meetings and commitment, incompatible information systems, conflicting objectives, independent cost evaluation of activities and processes, etc (Arshinder & Deshmukh, 2008) – this research provided evidence that SCC is also affected by the past experience of the organizations with suppliers and distributors and by institutions environment.

By analyzing the whole chain instead of just a dyad as the majority of SCC researches, this study contributes the theme debate by confirming the traditional barriers to coordination but also findings suggest that institutional and environment forces can act as barriers to an effective coordination of the supply chain. For managers, the paper provides a tool to identify what are the main aspects that should be surpassed to manage effectively processes and activities across organizations in a supply chain.

The research has some limitations. Firstly, the focus was given to multinational drug product reference industry. The research was limited to companies with operations mainly in the Southeast and South. No interview was conducted with companies with the highest performance in the North or Northeast. As the pharmaceutical market is highly regionalized, additional features should be found in these regions.

Moreover, because it is a case study, the results of this research are limited to companies that participated in the interviews, No generalization should be made.

REFERENCES

Amaral, G. L. (2006). *Radiografia da Tributação sobre Medicamentos: Carga Tributária Incidente no Setor Farmacêutico.* São Paulo: Federação Brasileira da Indústria Farmacêutica (Febrafarma).

Arshinder, A. K.& Deshmukh, S.G., (2008) Supply chain coordination: Perspectives, empirical studies and research directions, *International Journal of Production Economics*, v. 115, p. 316-335.

ASSOCIAÇÃO BRASILEIRA DA INDÚSTRIA QUÍMICA (ABIQUIM). (2014) *A Indústria Química – Estatísticas.* http://www.abiquim.org.br/conteudo.asp. Accessed in November 04th.

Barratt, M., Choi, T. Y., & Li, M. (2011). Qualitative case studies in operations management: trends, research outcomes, and future research implications. *Journal of Operations Management*, 29(4), 329-342.

Bastos, V. D. (2005) Inovação Farmacêutica: Padrão Setorial e Perspectivas para o Caso Brasileiro. *BNDES Setorial*, n. 22, p. 271-296.

Bowman, C. & Ambrosini, V. (2000). Value Creation Versus Value Capture: Towards a Coherent Definition of Value in Strategy. *British Journal of Management*, 11(1), 1-15.

Brandenburger, A. M., & Stuart, H. W. (1996). Value-based Business Strategy. *Journal of Economics & Management Strategy*, 5(1), 5-24.

Capanema, L. X. L. (2006) A Indústria Farmacêutica Brasileira e a Atuação do BNDES. *BNDES Setorial*, n. 23, p. 193-216.

Chen, I. J.& Paulraj, A. (2004) Towards a Theory of Supply Chain Management: The Constructs and Measurements. *Journal of Operations Management*, v. 22, n. 2, p. 119-150.

Coff, R. W. (1999). When competitive advantage doesn't lead to performance: The resource-based view and stakeholder bargaining power. *Organization Science*, 10(2), 119-133.

Cooper, M. C.& Ellram, L. M. (1993) Characteristics of Supply Chain Management and the Implications for Purchasing and Logistics Strategy. *The International Journal of Logistics Management*, v. 4, n. 2, p. 13-24..

Cooper, M. C.; Lambert, D. M. & Pagh, J. D. (1997) Supply Chain Management: More Than a New Name for Logistics. *The International Journal of Logistics Management*, v. 8, n. 1, p. 1-14.

Crook, T. R., & Combs, J. G. (2007). Sources and consequences of bargaining power in supply chains. *Journal of Operations Management*, 25(2), 546-555. doi: 10.1016/j.jom.2006.05.008

Croxton, K. L. et al. (2002) The Demand Management Process. *The International Journal of Logistics Management*, v. 13, n. 2, p. 51-66, 2002.

Forrester, J. W. Industrial Dynamics. Cambridge, Mass: Massachusetts Institute of Technology, 1961.

Fugate, B., Sahin, F. & Mentzer, J.T., (2009) Supply chain coordination mechanisms, *Journal of Business Logistics*, v. 27 (2), p. 129-161.

Geary, S.; Disney, S.M.& Towill, D.R. On bullwhip in supply chains – historical review, present practice and expected future impact. *International Journal of Production Economics*, v. 101, p. 2-18, 2006.

Holweg, M.; Disney, S.; Holström, J.& Smaros, J. Supply Chain Collaboration: Making Sense of the Strategy Continuum. *European Management Journal*, v. 23, n. 2, p. 170-181, 2005.

Ketokivi, M.& Choi, T. (2014) Renaissance of case research as a scientific method. *Journal of Operations Management*, v. 32, p. 232-240.

Kinra, A.& Kotzab, H. (2008) A macro-institutional perspective on supply chain environmental complexity. *International Journal of Production Economics*, v.115, p. 283-295.

Lambert, D.; Cooper, M. C.& Pagh J. D. (1998) Supply Chain Management: Implementation Issues and Research Opportunities. *The International Journal of Logistics Management*, v. 9, n. 2, p. 1-19.

Lee, H. L.; Padmananhan, V.& Whang, S. (1997) The Bullwhip Effect in Supply Chains. *Sloan Management Review,* p. 93-102, Spring.

Li, X. & Wang, Q. (2007) Coordination Mechanisms of supply chian systems, *European Journal of Operational Research*, v. 179, p. 1-16.

Magalhães, J.L; Boechat, N. & Antunes, A.M.S.(2008) An overview of the Brazilian pharmaceutical production status. *Chemistry Today*, v. 26 (4).

Mentzer, J. T. et al. (2001) Defining Supply Chain Management. *Journal of Business Logistics*, v. 22, n. 2, p. 1-25.

Miguel, P.L.S.; Reis, M.A.& Pignanelli, A. (2009) Gestão da Demanda em Cadeias Farmacêuticas Brasileiras: Um estudo de Casos Múltiplos. In Encontro de Anpad, 2009, São Paulo. XXXIII Encontro da Anpad.

Romano, L. A. N.; Pelajo, M. A.& Silva, M. A. C. (2007) *Análise de Desempenho Econômico-*

Financeiro do Setor Farmacêutico no Brasil: 2003 a 2005. São Paulo: Federação Brasileira da Indústria Farmacêutica (Febrafarma).

Sahin, F.& Robinson, E.P. (2002) Flow Coordination and Information Sharing in Supply Chains: Review, Implications and Directions for Future Research, *Decision Scineces*, v. 13 (4), p. 1-32.

Santos, J.B.& D'Antone, S.A. (2014) Reinventing the wheel? A critical view of demand-chain management., *Industrial Marketing Management*, v. 43, p. 1012-1025.

Tang, O., Cao D.& Schvaneveldt, S. J. (2008) Editorial Institutional perspectives on supply chain management. *International Journal of Production Economics*, v.115,p. 261.

Taylor, D. H. (2006) Demand Management in Agri-Food Supply Chains: An Analysis of the

Characteristics and Problems and a Framework for Improvement. *The International Journal of Logistics Management*, v. 17, n. 2, p. 163-186.

Xu, L.& Beamon, B.M. (2006). Supply chain coordination and cooperation mechanisms: An attribute-based approach, *The Journal of Supply Chain Management*, Winter, p. 4-12.

Yaibuathet, K; Enkawa, T.& Suzuki, S. (2008)Influences of institutional environment toward the development of supply chain management. *International Journal of Production Economics*, v.115,p. 262-271.

Yin, R. K. *Case Study Research: Design and Methods* (2nd ed.), Newburry Park: Sage Publications, 1984

APPENDIX I – INTERVIEW PROTOCOL

Name of interviewer: Organization:

Function: Since:

Section 1 - Internal and intra-organizational relationships

1. Could you explain your role in the relationship between your organization and other companies in the chain?

2. In your perspective, how was the historical evolution of your organization's echelon in the Brazilian pharmaceutical supply chain?

3. In your opinion, which is the most powerful echelon in this supply chain? Are companies dependent of each other?

4. How do you describe the chain in terms of commercial issues? How are the price and conflict negotiated with customers? And what about suppliers?

5. Could you explain how are the internal relationships in your organization? Do different departments share the same objectives and vision?

6. How do you define the intra-organizational relationships in the chain? Do you describe them as collaborative or competitive?

7. The pharmaceutical supply chain is known as very profitable. How is this profit shared between organizations in the chain?

Section 2: Internal process of each organization.

8. Who is responsible for demand and supply management in your organization? Are demand and supply synchronized in your company? How?

9. What are the mains sources of inventory in this organization? What are the key issues? How does demand variability impact inventory management?

10. Could you briefly describe incentives and discounts programs adopted by this company? Do these programs impact the demand profile?

11. How does the government regulation affect your demand management process (consider Anvisa, price control, fiscal benefits)?

12. Is there any initiative for technology integration in the chain? With whom? Which are the objectives?

13. In your opinion, how could the demand management process be improved? Would this process improvement have impact performance?

APPENDIX II – LIST OF CODES

Goals and Objectives Conflicts (GOC)

Leadership (L)

Different perspectives on decision making (DP)

Organizational culture (OC)

Unique performance measurement (PM)

Coherent contracts (CC)

Information sharing (IS)

Information systems (IS)

Operational practices (OP)

Internal synchronization (IS)

Intraorganizational integration (II)

Bargaining power (BP)

Dependency (D)

Government influence (GI)

Supply chain history (SCH)

AUTHOR'S BIOGRAPHY:

Priscila Laczynski de Souza Miguel : has graduated in Chemical Engineer by Unicamp and has done her Master and PhD in Business Administration in FGV-EAESP. Currently she is a Lecturer at Operations Management Department at FGV-EAESP and is researcher at the Center of Excellence in Logistics and Supply Chain FGV (GVCELog). Her interests are: Logistics, Supply Chain Management, Buyer-Supplier Relationships and Operations in Natural Disasters.

Manoel A. S. Reis: Holds Bachelor and Master degrees in Naval Architecture from the Polytechnic School of the University of Sao Paulo, Brazil, and a Ph.D degree from the Massachusetts Institute of Technology – MIT – USA.

At Getulio Vargas Foundation in Sao Paulo, Dr. Reis is a Professor of Logistics and Supply Chain, Director of the Center for Logistics and Supply Chain Studies (GVcelog), and Coordinator of the courses Master in Logistics and Supply Chain and continuing education Business Logistics.

10

Supply chain governance in the production systems of business clusters: The case of the footwear industry in Jaú

Eliane Pereira Zamith Brito
eliane.brito@fgv.br
FGV-EAESP

Lilian Soares Pereira Carvalho
lilianpereiracarvalho@gmail.com
FGV-EAESP

ABSTRACT: This paper's objective was to analyze the role of transaction costs in the longevity of Jaú footwear supply chain. Firms are configured in a cluster, and their level of coordination was investigated. The study data was collected via conducting interviews with the cluster's agents, making direct observations during the field visits, and consulting secondary data emerging both from document analysis and from sources such as newspapers, sectoral studies, and previous academic studies. The data point to the intense frequency of transactions among firms and the low uncertainty of internal transactions within the cluster due to the extensive exchange of information among agents. The bonds of competition and coordination that are established among the firms both reduce transaction costs and make them more competitive. Both human asset specificity and site specificity are important factors in the longevity of Jaú's footwear cluster.

1. INTRODUCTION

Transaction cost economics is based on two behavioral assumptions: bounded rationality and opportunism. In transaction cost economics, economic agents are limitedly rational, so decisions are satisfactory but not optimal. This reasoning's use of cognition is bounded. Opportunism refers to the fact that human beings are not entirely reliable, and therefore, contractual safeguards are necessary when establishing transactions between economic agents. These behavioral assumptions are the generators of transaction costs (Farina, Azevedo & Saes, 1997, Williamson, 1985, 1996).

Transactions are described through three dimensions: frequency, uncertainty, and asset specificity (Williamson, 1985). Transaction cost economics proposes that the variety of contractual relationships among businesses can be explained by differences in the dimensions of transactions. For Williamson (1985,1996), contractual variety follows a continuum from the perfectly competitive market to the vertically integrated firm, which that author calls hierarchy, passing through intermediary forms of business coordination, including business networks, franchises, and *joint ventures*. This work investigated a specific form of contractual relations among firms in the footwear supply chain of the Jaú cluster.

Cassiolato and Lastres (2003) understand clusters as a concentration of firms both in the same supply chain and in the same location. According to other researchers, a cluster is not only a concentration of businesses in the same location but also a network of linked firms (Becattini, 1999, Perry, 1999, Robertson & Langlois, 1995). Thus, clusters can be analyzed using transaction cost economics: more specifically, it is possible to investigate both participants businesses' level of coordination and how the dimensions of transactions are established among supply chain members.

The concept of a business cluster was brought to the attention of economists very early in the history of economy, by Alfred Marshall (1891) and experienced an academic "rebirth" with studies on the Third Italy, i.e., the Italian clusters that according to Becattini (1999) were almost entirely responsible for the positive results of Italian exports after the Second World War. Notwithstanding, what are the reasons that small businesses have the ability to both compete individually with large global firms and achieve good results such as those achieved by

the abovementioned Italian exports? For Becattini (1999), competition and cooperation bonds between firms both reduce transaction costs and make them more competitive. Additionally, Becattini (1999) and Suzigan et al. (2003) assert that transaction costs impact the longevity and the competitiveness of clusters' supply chains, however, those authors did not empirically investigate these relationships. Therefore, this article describes not only transactions in the cluster's supply chain but also the influence of those transactions on the cluster's longevity.

The latest data from the Brazilian Institute of Geography and Statistics (Instituto Brasileiro de Geografia e Estatística—IBGE) on industrial activity (PIA, 2011) indicate that footwear production employs slightly more than 4% of Brazil's industrial workforce and represented 2% of the Brazilian footwear sales in 2001 (Côrrea, 2001). In 2013 and 2014, the State of São Paulo created more jobs in the footwear industry than all but two of Brazil's states (ABICALÇADOS, 2014). The Jaú, Franca, and Birigui regions produce the most footwear of any regions in the State of São Paulo, but Jaú is the only region to specialize in women's footwear. Applying Puga's (2003) cluster identification methodology, the Jaú region has systematically been identified as a cluster (post-1996 employment data from MTE/RAIS-CAGED were used). Moreover, the Federation of Industry of the State of São Paulo (Federação da Indústria do Estado de São Paulo) decided to promote the competitiveness of the Jaú region by support it as a footwear cluster. The activity of the footwear industry is one of the most important to the city of Jaú and the latest available estimates it represented 40% of the city of Jaú's gross domestic product (GDP), along with approximately 16,000 direct and indirect jobs (APL de Calçados de Jaú, 2007).

The study data were collected via interviews with the cluster's agents, direct observation during field visits, and secondary data emerging from document analysis and from sources such as newspapers, sectoral studies conducted by the National Bank for Economic and Social Development (Banco Nacional de Desenvolvimento Econômico e Social—BNDES), and previous academic studies.

In addition to this introduction, this article presents the theoretical framework for the concepts essential to developing the research, the research procedures, and the study results. The final section contains an analysis of this work's theoretical contributions.

2. LITERATURE REVIEW

In his pioneering article *"The nature of the firm,"* Coase (1937) seeks to break with the neoclassical economic paradigm in which firms are seen either as "black boxes" or in terms of their production functions. According to this economic perspective, the market is composed of several buyers and suppliers with irrelevant identities, and price is sufficient to communicate all of the necessary information about a product or a service. Thus, Coase aimed to understand what the firm is and why it exists, describing its scope, extent, and limits. He sought to identify the reasons that some resources can be acquired in the market by price mechanisms, whereas other resources are produced by an entrepreneur's coordination mechanism, i.e., within the limits of firms.

Zylbersztajn (1995) states that Coase recognized that markets do not operate with zero transaction costs. That is, for Coase, market transactions create costs and a firm formation overseen by an authority—i.e., the entrepreneur, who manages the use of resources, brings the benefit of saving the costs associated with market transactions.

The primary explanation of why it is lucrative to establish a firm is the existence of costs through the use of the price mechanism, i.e., acquiring market resources. The most obvious market cost of resource acquisition is associated with discovering the best price. Other costs that should be considered are those of negotiating and of preparing a contract (Coase, 1937).

Coase (1937) also asks the following question: if costs of the use of the price mechanism exist, why do market transactions exist? The response lies in firms' growth limits. First, as firms grow, there may be diminishing returns for the entrepreneur's use of coordination. Coase explains that there is a point at which the cost of organizing an additional transition within a firm is equal to the cost of the market transaction. Second, with the firm's growth and the increase in the number of functions and internalized transactions, the entrepreneur cannot make the best use of production factors. Analogously, there is a point at which the resource-waste cost equals the market transaction cost. These two reasons correspond to the "diminishing returns to management," a phrase used by economists (Coase, 1937, p. 12).

Williamson (1985, 1996) is an exponent of transaction cost economics, and his primary contribution has been to expand Coase's (1937) thinking and transform it into a coherent theoretical corpus. Thus, that author considers the transaction as a unit of analysis, assigning dimensions to transactions, so that the difficulties of transaction-cost observation and measurement can be overcome. To explain the existence of transaction costs, Williamson (1985) bases his theory on two behavioral assumptions which are bounded rationality and opportunity. The existence of transaction costs limits the efficiency of the structures that govern (i.e., coordinate) these transactions. In view of these transactions' dimensions—defined by Williamson (1985) as frequency, uncertainty, and asset specificity—the structures of governance will have different levels of efficiency. The author defines three structures of governance: the market, hierarchy, and the hybrid structure.

For Williamson (1985), the three structures of governance are institutions of the capitalist system whose primary object is to minimize transaction costs. These costs are managed more efficiently by assigning structures of governance to the transactions, those structures differ in their capacities for adaptation and the costs associated with them.

The theory of transaction costs considers the problem of economic organization as a contractual problem. In this context, the firm is viewed as a complex (*nexus*) of contracts. These contracts can be either explicit (i.e., formal) or implicit (i.e., informal). Each type of contract is associated with a different support apparatus, but both are subject to transaction costs, which can be *ex-ante* or *ex-post* (Williamson, 1985). *Ex-ante* transaction costs are those that occur prior to the transaction (Kreps, 1990), and they can be the costs of preparing, safeguarding, and negotiating a contract. The contract may be drafted relatively comprehensively, anticipating a number of contingencies, but it will never be fully prepared because of bounded rationality, and the parties must resolve contingencies as they occur. Contractual safeguards can have innumerable forms, the most typical is common ownership. To avoid transactions with various agents, a firm can choose to internalize an activity, substituting the market by an internal organization, or it can even establish a long-term contract with an agent. Negotiating a contract refers to the costs that occur when agents define and safeguard their interests, so that there is a dispute among the parties concerning planned or unplanned contingencies.

In turn, *ex-post* transaction costs are tied to the execution and continuity of the transaction (Kreps,

1990). These costs are divided into the following categories: costs associated with poor adjustment when transactions or their specifications change at any given time, costs associated with adjustments to correct contractual specifications, costs associated with the structures of government (or the use of the legal system) to resolve contractual disputes, and costs associated with monitoring to ensure the parties' commitment (Williamson, 1985, p. 21). Thus, a contract may stipulate x, but after some event, the parties decide that y should be done. Changing from x to y may not be an easy task. This change will initiate a process of renegotiating the contract, which gives rise to the emergence of opportunistic behavior. If the dispute is not amicably resolved, the legal system (courts, courts of arbitration, etc.) should be used, which entails greater transaction costs (Williamson, 1985).

Given that the use of the legal system is onerous, it can be inferred that the economic system's efficiency is limited by the set of institutions that regulate this system (Farina et al., 1997). According to North (1990), the institutions are the rules of the game, whether it is economic, political, or social. An example of the "rules of the game" can be a country's laws and its legal system, which is the institution that applies those laws.

In that notion lies the complementary nature of the analytical levels of transaction cost economics. North's (1990) "game rules" include the analytical level of the institutional environment. The structures of governance (market, hybrid, and hierarchy), which is another of the analytical levels proposed by Williamson (1985), are subject to the rules of the institutional environment, but can act strategically (as in an industry *lobby*) to change them. The final analytical level is composed of the individuals, whose behavioral attributes of bounded rationality and of opportunism act essentially on the choices and the preferences that influence the structures of governance and in the world of contacts among economic agents.

Despite the existence of three analytical levels for the study of transaction costs, this work will focus on the structures of governance, which is the same focus as the works of Williamson (1985, 1996).

3. RESEARCH METHODS

The case study method was employed here. The object of the study was Jaú's footwear cluster, which is composed of various agents such as footwear firms, stitching centers, and support organizations such as colleges, technical schools, trade unions, and Sebrae (Brazilian Service for Support to Micro and Small Businesses of the State of São Paulo). Therefore, both various local actors that participate in the region's footwear supply chain and local agents familiar with the production systems were consulted. Much of the data collected, primarily during interviews, revealed informal aspects of the relationships among agents, enriching the description of those relationships, as Godoy (2006) suggests as a possibility in the case study approach.

For this study, three data sources were used: interviews with the cluster's agents, direct observations during field visits, and secondary data derived from documentary analysis and from sources such as newspapers, sectoral studies performed by BNDES, and previous academic studies.

According to Merriam (2002), the quality of research is related to the production of valid and ethically reliable knowledge and the author proposes to use triangulation to ensure validity and reliability. Four types of triangulation can be used to confirm a study's conclusions: multiple researchers, multiple theories, multiple data sources, or multiple methods of analysis. In this study, triangulation was achieved using multiple data sources.

With respect to secondary data, sources such as regional newspapers and previous studies were used. For the latter, we used academic journals such as Revista de Administração de Empresas (RAE), the proceedings of national conferences, and theses in the libraries of the State University of Campinas (Universidade Estadual de Campinas—Unicamp) and the University of São Paulo (Universidade de São Paulo—USP). We identified one dissertation with focus on the Jaú cluster, it was authored by Oiko (2007), of the USP's São Carlos School of Engineering. We also used another study on this cluster, which was sponsored by representatives of the industry association and of Sebrae and was performed by Oliveira and Garcia (2001), of the School of Geography of São Paulo State University (Universidade Estadual Paulista—UNESP).

We also analyzed documents discovered in the archives of Jaú's footwear industry association. They contained some editions of the informational journal distributed to industry association members, folders and pamphlets about the industry association's action plans, and documents with information about industry association members. These data provided data about the relationships that the firms established in the clusters, including but not limited to information about whether businesses act together to participate in (national or international) trade shows, whether businesspeople meet frequently to exchange information, and whether there is skilled manpower in the sectoral associations.

Field notes from observations performed during three visits to the cluster were used as a primary data source. We made observations during the interviews with firms and other agents, in addition to visits to the Jaú Footwear Industry association (Sindicalçados Jaú), the local office of the National Service for Industrial Training (Senai, 2014), the city-donated industrial districts, and the city's three footwear malls. The observations were performed informally. The data from these observations were compiled in field notes.

We conducted in-depth interviews and we used them as a primary data source. This approach was used to assess the perceptions of the cluster's agents concerning the dimensions of transactions and their influence on both transaction costs and the cluster's longevity. We used an interview script to guide during the interviews. The following subjects were interviewed: two businessmen in the footwear industry, two stitching-center proprietors, a representative of the employers association, a representative of the local Sebrae office, a local professor, a Senai consultant to the stitching centers, a representative of Senai's design center, and a Senai teacher.

We analyzed the data obtained using content analysis technique. We sought to identify analytical categories related to the theoretical framework. Four primary categories were established: the three dimensions of transactions as defined by Williamson (1985, 1996)—i.e., frequency, uncertainty, and asset specificity, and a fourth analyzed category of the cluster's longevity. These categories were divided into subcategories for better data clustering.

The frequency category was divided into three subcategories: joint action, reputation established between the agents, and the existence of credible commitments. For the purpose of analyzing the frequency dimension, joint action was established as a subcategory because it can be treated as a *proxy* of transaction frequency. These transactions are not necessarily buying and selling relationships exclu-

sively, but also interorganizational and interpersonal relationships that may, over time, affect transaction costs. For instance, the interviewees were asked whether they traveled with people from other raw materials or finished-goods firms. This question was asked to determine whether there are any long-term relationships among the studied agents. This choice of joint action as a subcategory of the frequency dimension is justified in that transactions between firms or between members of the same community can reoccur in such a way that constructing a complex mechanism of governing these transactions can become economically feasible.

In a cluster, which this study treats as a structure of complex governance analogous to networks, frequent transactions can also occur between community members, to the extent that they establish relationships with other agents at the same level in the supply chain for purposes other than buying and selling in the vertical direction of the supply chain. These recurrent relationships can both influence the construction of reputation and reduce transaction costs. Following this reasoning, the second subcategory we analyzed was the establishment of reputation among agents. As economic agents transact with each other, they establish a reputation as the result of more-frequent transactions (Farina et al., 1997, Williamson, 1985). Finally, according to Williamson (1985), credible commitments can safeguard a relationship from opportunist behavior. According to that author, credible commitments arise when a transaction presents the prospect of continuity. Thus, when transaction continuity is expected, opportunistic behavior today can undermine tomorrow's gains.

The uncertainty category has three subcategories: the exchange of information between agents, social sanctions, and access to market information without a complex collection apparatus. The exchange of information among agents is defined as a subcategory in accordance with Dyer (1997), who argues that a greater level of information exchange among agents reduces uncertainty. Moreover, agents can exchange information about the reputation of suppliers or customers, thus establishing social sanctions. An agent can convey information about a customer or supplier's opportunistic behavior, and thus, the cluster's agents know ahead of time about that opportunistic behavior and choose not to transact with the opportunist. Dyer (1997) also notes the existence of social sanctions, which decreased the incidence of opportunistic behavior. For Dyer (1997), these sanctions occurred such that suppliers would lose future contracts. If an agent in the chain behaves opportunistically, it is not contractual safeguards that are invoked, but social sanctions, in the form of either reduced orders or the termination of the relationship. Finally, access to market information without a complex collection apparatus reflects the fact that the businesses in a cluster do not always need to establish marketing, design, and other departments to obtain market information. That is, they do not necessarily need to internalize these functions because they obtain the information from support agencies or from other firms. According to Williamson (1985, 1996), strong administrative controls are necessary for those functions essential to the firm. In a cluster, if there is an intense exchange of information, these functions are "in the air," as stated by Marshall (1891). Therefore, firms do not need to internalize the search for market information if it can be obtained from other agents. That is, firms can use support institutions to access consumer market research about trends in footwear design, etc. Thus, uncertainty can be lessened without the costs associated with a more hierarchical structure.

The asset specificity category was subdivided according to the six types of asset specificity defined by Williamson (1985, 1996): site specificity, physical asset specificity, human asset specificity, brand names, temporal specificity, and dedicated assets.

The final category analyzed was the longevity of the cluster. Jaú's footwear cluster was classified as a cluster by the methodology of Puga (2003) for every year analyzed (since 1996). This study sought to determine whether the interviewed agents perceive advantages in establishing themselves as a cluster and the difficulties of leaving their current location, seeking to identify the reasons that the cluster is maintained.

Despite the relevance of the demand of national market to increase the competitiveness of local industries (Porter, 1991), this analysis focused on the offer-side of the footwear industry.

Due to space limitations, it was not possible to include interview excerpts, observations, or other evidence in this article.

4. RESULTS

The footwear supply chain is presented in the Figure 1.

Figure 1 - Footwear supply chain

Source: Research data

The first category analyzed was frequency. This category was divided into subcategories, the first being joint action between businesses. The interviewed professor explains that firms began to act in a coordinated manner when a project that emerged from the need of disposing industrial waste was implemented. This project occurred because local firms had to comply to the waste legislation and together they could do it cheaper than by itself. The leather and fur industry, which includes the footwear industry, is among the highest polluter economic activities, according to a survey performed with data from 1996 (CETESB, 2014). Thus, according to the interviewee, the contacts among businesspeople became more

cohesive, and a greater understanding emerged regarding the advantages for all if the businesses were to interact more and to think more about a regional strategy. Even so, according to those interviewed businesspeople, this work is largely due to the connections established by local institutions.

One of the businesspeople interviewed states that the firms' joint activity is primarily performed through the coordination of local institutions, such as the Brazilian Service for Support to Micro and Small Businesses of the State of São Paulo (Serviço de apoio às micro e pequenas empresas do Estado de São Paulo—Sebrae) and the local industry association, to develop exports, to strengthen the cluster's sales, and to provide access to public resources, such as city-donated land and lower taxes. Another interviewed businesswoman asserts that the most cohesive work is recent, as indicated in this passage: *"I think that such work already exists, of working together, but I think this work is asserting itself now, during the last year or so. The trend of working together, of the clustered firms is increasing."*

For the representative of the local Sebrae office, the joint action of the firms does not occur in a systematic manner, despite institutional efforts, but it did create both cohesion and practical benefits for solutions to regional businesses' common problems.

In addition to the perception that joint action can bring benefits for everyone, another of the region's businessmen highlights more practical results of institutions' activities, primarily those of the industry association.

Fairs are opportunities to promote Jaú's firms to potential customers; the industry association, the local government, and Sebrae all play central roles in organizing and encouraging business people's joint participation. Every July, with support from Sebrae, local footwear manufacturers participate in the joint Jaú stand at Francal (Sindicalçados, 2014), the largest trade and fashion fair in Latin America's footwear and accessories sector, which attracts both national and international buyers. Footwear manufacturers from Jaú also participate in the Couromoda at a joint stand subsidized by Sebrae and the local government. Couromoda is also an important trade and fashion fair to the sector. In 2014 Couromoda had more than 35 thousand visitors, 1500 exhibitors and buyers from 66 countries (Couromoda, 2014).

The Jaú footwear industry association is another important local agent that promotes the collaboration between its associated firms.

Finally, it can be emphasized that there is political activity on behalf of the region, including an elected federal deputy who works not only for the interests of Jaú but also for other footwear clusters. The industry association works side-by-side with that legislator.

The second category to be analyzed is the establishment of agents' reputations. This subcategory is directly influenced the exchange of information category, because it is unnecessary for an agent transact directly with another agent to know his or her reputation—it is enough that a cluster agent transfers

information about how the supplier or customer behaves so that a positive or negative reputation is transmitted. In the event of prior opportunistic behavior, a social sanction is established: other agents of the cluster will no longer do business with an agent who has behaved opportunistically.

The interviewed businesspeople's choice of suppliers is influenced by the reputation of the region's agents. One of the interviewed businesspeople states that he has a long-term relationship with some of the cluster's suppliers and that through that relationship he has established a reputation based both on his experience as businessman and on repeated transactions over time.

For the Sebrae representative, the establishment of reputation is so important that in that representative's perception, there is a search for higher-quality suppliers outside the cluster as the region's businesspeople are becoming increasingly professional.

Stitching centers are an important link in the Jaú footwear chain. Reputation establishment is also a reality among these service providers: firms recognize the centers' reputations and centers recognize the firms' reputations. For the stitching-center proprietors interviewed, company recognition is significant and influences the decision of whether is worthwhile to work with a particular company. The owner of one of the stitching centers, which is established on a commercial property, states that he always works with the same company unless there is something that prevents him from doing so, such as opportunistic behavior or noncompliance with an agreement.

For the interviewee from another stitching center (located in his place of residence), the long-term relationship and establishment of reputation were enabled by the fact that the frequency of transactions was constant even in times of low demand, and the knowledge of the business partner for which he worked for two years (of the four years that his stitching center was established in the city).

The businesspeople interviewed stated that their firms maintain long-term relationships both with local suppliers and with customers, the latter of which are primarily located outside the region. For the professor, switching suppliers only occurs if there is a large advantage in price or the timeliness of delivery, but for him, the greatest influence would be fashion (i.e., the design of the footwear). If a supplier is not current with the current collection's design, then switching suppliers will be more frequent. So, suppliers have to develop new capabilities to meet market demands.

According to the consultant to the stitching centers, 90% of the Jaú centers maintain long-term relations with firms. The owner of the stitching center at the commercial property states as follows: *"I know people that work 12 years with the same firm, 13 years. Me, the longest I stayed with a firm was five years."* Another stitching-center owner interviewed makes the same point, stating that the firm with which he has worked for two years treats all of the stitching centers equally and keeps its word in the sense that it continues to provide the center with work, even if only to create inventory. This observation shows the importance of one's "word" in the business relationship with the company.

In summary, the frequency category analyzed through the subcategories of joint action, the reputation established among agents, and the existence of credible commitments demonstrates that the following:

- With respect to joint action, the agents initially established relationships because the need to dispose solid waste from the production process and since then, the connections of support organizations (such as the trade industry association and Sebrae) has enabled the expansion of export sales and sales to the national market, such as, for example, with the fairs and the collective stands, in addition both to access to public resources and workface training in conjunction with Senai (National Industrial Training Service),

- Reputation is important to and influences the choice of suppliers, and

- Credible commitments appear to be of great importance for the local cluster's agents, given the long-term relationships established among agents.

The second category analyzed was that of uncertainty, as explained above. This category was divided into three subcategories: the exchange of information between agents, social sanctions, and access to market information.

Through the discourses of the interviewees, one can perceive that the exchange of information among agents is intense. All of the interviewees confirmed the existence of the exchange of information among agents, but always in an informal way, via "chit-chat."

Another interviewed businessman states that the exchange of information with other cluster agents is primarily related to suppliers. Regarding the search for new suppliers, the Sebrae representative says that when agents seek another supplier, they first seek to learn about the supplier's behavior, as indicated by the following statement: *"(...) For a new relationship he will do this, he will seek another type of information to determine whether it is meritorious, whether the supplier has a pedigree, something along those lines"* (Sebrae representative).

Thus, one perceives the influence of the exchange of information on the dissemination of an agent's reputation. Reputation is established for agents with which one transacts and through the exchange of information; an agent's propensity to engage in opportunistic behavior can be discovered in advance. In the interviewee's words, it is important to discover whether a supplier has a "pedigree", i.e., if the supplier has a positive reputation that enables an agent to consider whether it will transact with him or her. Stitching-center proprietors likewise affirm that they exchange information among themselves and demonstrate the second analyzed subcategory, the existence of social sanctions.

The interviewees demonstrate the informal exchange of information and report that this exchange exists on all levels, namely, the levels of suppliers, customers, and company employees. The aforementioned statement from the Sebrae representative stresses that by exchanging information about employees, businesspeople seek references about a prospective employee and if he or she has a negative reputation, then he or she will not be hired, which indicates that social sanctions exist even for employees, thus greatly reducing uncertainty in labor relations.

The final analyzed subcategory is access to market information. The businesspeople interviewed emphasize the actions of institutions, primarily Sebrae and the industry association, related to gaining access to market information. The interviewees state that footwear design and fashion trends are based on customer – the retailers - requests. Accordingly, one of the interviewees emphasizes that clients dictate also the sale of footwear, specially because they build the brands.

In the document analysis, the actions promoted by the industry association and Senai partnership to provide Jaú businesspeople with access to design information also became evident. Senai organize workshops to provide guidelines for creating footwear for new collections, which are important to the individual firm and also to the cluster. Finally, institutions showed a clear interest in inserting Jaú firms into the Brazilian fashion circuit by promoting, through a partnership among Sebrae, the industry association, the Brazilian Association of Fashion Designers (ABEST), and the Brazilian Association of Firms of Components for Leather, Footwear and Manufactured Goods (Assintecal).

In sum, uncertainty in the Jaú footwear cluster's transactions appears to decrease to the extent that

- Agents informally exchange information among themselves about suppliers, customers, and employees,

- Social sanctions are established for agents with a poor reputation, to the extent that the exchange of information about these actions circulates among cluster agents,

- Agents have access to market information without establishing research departments for this purpose and they not only use customers (retailers) to understand fashions and design but also use support institutions both to understand fashion trends and to obtain information about the consumer market.

The third category analyzed was asset specificity. It aimed to identify the types of asset specificities that are both present and relevant to the cluster's longevity. The asset specificity was divided into six subcategories: site; human assets; physical assets; dedicated assets; brand; and temporal.

The site asset specificity refers to the potential cost reductions that the proximity of the same supply chain firms can obtain. The proximity of the footwear production chain firms in Jaú promoted a division of labor in order to improve productivity. The list of firm members of the industry association shows that - footwear manufacturers, components manufacturers (soles, clapping, buckles, heels, and others), cartons and packaging producers, and tanneries. Probably not all firms in the region are listed, but this information provides an estimate of how productive chain processes are present in the cluster, benefiting individual firms by geographical proximity.

The interviewees perceive no joint action among firms of the cluster but it was possible to identify collaboration among small groups of firms. Furthermore, there are three malls composed exclusively with footwear stores. These malls have 175 footwear stores and to establish them, regional footwear manufacturers worked together to maximize their earnings.

The second subcategory analyzed was human asset specificity. There are two fundamental activities to the efficiency of women footwear industry – cutting and stitching. A relationship between site specificity and human asset specificity can be established in the case of stitching centers. One of the stitching-center proprietors was not originally from Jaú or even from the footwear industry. In other words, Jaú is a footwear hub in which the outsourcing of labor is so intense, the interviewee can establish himself in the city and learn the stitching technique, something that would not be possible in another location.

Evidence was found of the need for strong administrative control of specific human assets, represented here by the figure of the cutter. The cutting phase is very important to have a good quality footwear and also to have lower raw material costs. An experienced cutter can improve final product quality and also minimize leather waste. An interviewee compares the amount of labor required during the footwear stitching and cutting phases to justify why cutting is primarily internal to firms, whereas stitching is external. Few leather cutters are necessary to have a large daily production. Thus, a cutter's labor is less expensive to the company than a stitcher's labor, not in terms of individual wage but in terms of the total final cost of the footwear. Other reasons identified by the interviewee for outsourcing the stitching phase involve logistical facilities, now that the stitching centers are all located within the cluster, in addition to the actual fixed costs related to the increased physical space required by the company in the event of the stitching phase being internalized. The interviewee also emphasizes the importance of avoiding labor costs by hiring outsourced stitching services.

It can be perceived that one very important factor in the longevity of Jaú's footwear cluster is the availability of specific human assets. Secondary data from a study conducted in 2001 support this statement and confirm the interviewees' discourse. According to Oliveira and Garcia (2001), in a quantitative study performed on the Jaú cluster between

1997 and 1998, 62 of the region's footwear firms were interviewed and one of the questions referred to the primary advantages of the Jaú location. Twenty-two of the interviewees responded that the Jaú location's primary advantage is the availability of a specialized workforce. That is, for approximately 35.5% of the firms interviewed, the primary advantage of being located in the cluster is the availability of a specialized workforce. It is worth noting that of all of the studied advantages, this was the one that was noted with the greatest frequency, especially by respondents from small firms (i.e., those with less than 51 employees).

According to Sengenberger and Pike (1999), economic analysis usually either ignores the workforce factor or cites it as an afterthought. Thus, specific human assets are treated as not being an integral part of economic success. The authors assert that the presence of a specialized labor *pool* is a basic requirement for a cluster's success. They also claim that the skill of the workforce is channeled through the community, passing from parent to child, from colleague to colleague. Another possibility, according to the authors, is the workforce training provided by technical schools.

None of the interviewees attributed importance to dedicated assets, physical assets, or temporal specificity subcategories.

Brand's specificity is defined as non-physical and non-human capital embodied in a firm's name. This concept of "brand" can be extended to a location and its identification with a certain industrial activity. In this case, we can ask whether the Jaú cluster has a brand that represents a location that produces women's footwear. On its website, the city of Jaú presents the city as the "capital of women's footwear" (Nosso São Paulo, 2014). Regardless of the classification of the city as the women's footwear capital by both the local government and its own businesspeople, the interviewees confirm that there is an identification of the cluster with the product, i.e., women's footwear. For example, the interviewed local college professor explains that footwear in Jaú is not sold; it is bought. By this, he means that retailers or footwear sales representatives usually seek out footwear factories and either purchase their production or suggest a design.

The interviewees were also asked whether the region's businesspeople are working jointly to build a cluster brand, including identifying their products

as produced in Jaú, similar to Birigui-São Paulo. The interviewees claimed that the responsibility for building this brand belongs to the Jaú industry association, but that there is awareness of the issue among the region's businesspeople. In the industry association's plan of action, the following goal was established: "The Jaú hub will be known as a fashion reference point in innovative and desirable women's footwear and accessories."

In sum, the analyzed asset specificity category showed the following:

- Regarding the subcategories of physical asset specificity, dedicated assets, and temporal specificity, the interviewees did not attribute importance to these categories. For them, physical assets are easily re-employable, there is no investment in assets performed specifically for an agent with which they transact, and there is no decrease in the transaction's value over time,

- Site specificity is shown through the presence of various agents of the cluster's supply chain and through the footwear factory stores, highlighting the importance of the footwear industry to the regional economy,

- Human asset specificity is important for the interviewees and is an important factor for the presence of the firms in the cluster. A specific workforce is available that in the opinion of the interviewees, has a handmade character and lacks professionalization.

- The location is associated with the production of women's footwear, but the identification of local products with an established brand for either the retailer or end consumer has not yet been achieved. According to the interviewees, building this brand is the responsibility of the support institutions.

The final category analyzed was the cluster's longevity. The question to determine longevity was the following: what are the primary difficulties of establishing and maintain a firm outside the cluster. For the businesspeople interviewed, the proximity of suppliers and qualified labor would be the primary difficulties associated with the distance from the cluster. Or in the words of an interviewed businessman, *"The current advantage is the workforce, proximity to the industry that adds to and generates activity for the association."* The local college professor cites the importance to the city of industrial footwear activity in a statement reminiscent of Marshall (1891), explaining that the secrets of the (local) profession are "in the air" and the interviewee claims that the local population "breathes footwear."

When the interviewee cites the importance of firms' professionalization, he wants to emphasize the handmade feature of women's footwear manufacturing, especially those made in Jaú. With respect to professionalization, he emphasizes both the difficult of having someone skilled to the job and the difficulty of working with various outsourced stitching centers. The interviewee means that this type of product requires a skilled workforce that would be very difficult to find elsewhere. The stitching center emerged through the time as a way to cope with rising indirect labor costs stitching center emerged through the time as a way to cope with rising indirect labor costs; which lead to collective use of outsourcing.

For the Sebrae and industry association representatives interviewed, the main difficulties for the firms in establishing themselves outside the cluster would be access to suppliers and a specialized workforce. The Senai teacher interviewed emphasizes the importance of the proximity of firms as a factor of the cluster's longevity. The stitching-center owners interviewed agree that it would be very difficult to establish themselves in a different location because of the need for firms in proximity. One owner interviewed says that for his center to leave the cluster, he would require a good offer with guarantees of continuity in the relationship.

In sum, with regard to the cluster's longevity category, the interviewees demonstrate the region's vocation for footwear production and emphasize the proximity of suppliers, a specialized workforce, and supply-chain agents as factors that would be difficult to find in another place, which in their opinion justifies the existence of the cluster. Considering the influence of transaction costs in the longevity of Jaú's footwear cluster, it can be perceived by the interviewees' statements that

- The primary reasons for the company to be located in Jaú are human asset specificity (specialized workforce) and site specificity (proximity to suppliers),

- Through the analysis of the above categories, access to information also decreases the uncertainty about relationships experienced by the

cluster's agents and can influence those relationships' continuity,

- With respect to the frequency category, credible commitments are also an important factor for the cluster's longevity because most of the firms establish informal agreements with both suppliers and customers.

5. CONCLUSIONS

Analysis of the collected data shows an intense frequency of transactions and low uncertainty for intermediaries of the supply chain because of the exchange of information among agents.

With regard to joint action, agents established inter-organizational relationships, in the first place, due to the need to dispose solid waste from the production process. Since then, joint actions of supporting organizations, such as industry association and Sebrae, have helped collective development in many fronts such as export trade, sales to the domestic market, collective stands in trade fairs, access to public resources but also in relation to the training of the workforce. The high frequency of inter-organizational and inter-personal interaction seems to indicate a reduction of transaction costs in the supply chain.

Reputation is important in the choice of suppliers, and those agents with good reputation are more likely to establish long-term relationships (Williamson, 1985, Farina et al, 1997). This may be the only way to differentiate vendors in this market of poorly differentiated suppliers of parts and services.

Respondents informed that supply chain members value the continuity of the transaction and they avoid breaking established relationships. The prospect of continuing the relationship leads to credible commitments (Williamson, 1985), so, opportunistic behavior has to be avoided, in order to increase the expectation of continuity of the relationship.

Supply chain members informally exchange information on suppliers, customers and even employees. We can see here that agents pass on information about the reputation of others. This exchange of information between agents decreases the uncertainty (Dyer, 1997) and therefore may reduce transaction costs. It is also used to establish social sanctions (Dyer, 1997), especially for agents with a bad reputation, to the extent that the exchange of information on these actions circulates among cluster agents.

The interaction among supply chain members and also with supporting institutions allows many suppliers to access market information without establishing research departments. Thus, the agents acting in the cluster have access to information about fashion, design, technical knowledge, and so on. In a cluster, if there is intense exchange of information, firms do not need to internalize the information search (Asheim, 2000).

The spatial proximity of firms in the same supply chain means lower costs for the chain (Williamson, 1985; Farina et al., 1997; Dyer, 1997). This locational specificity is well established in the cluster both in the production side and in the sales side.

The human asset specificity is probably a main explanation for the presence of companies in the region. But the specific skilled workforce available in the cluster lacks professionalism.

Firms in the cluster usually do not invest in assets that can lock them in with other firm, so they can easily reallocate their assets without losing value.

The city presents itself as a producer of women's footwear and is recognized by regional and national market as a major producer of this type of product. However, the identification of local products with an established brand to be used at retail level, i.e., to create consumers' preference, is not yet clear. Supply chain members interviewed understand local supporting institutions and government as responsible for building the local brand for footwear manufactured locally. Despite that, manufacturers whose products have clear differentiation invest in the development of their brand.

In sum, cluster's longevity is due to the existence of raw material suppliers, skilled labor, other members of the supply chain located in the region, and also because these factors are difficult to find together elsewhere. The agglomeration also promotes innovation, that is another important factor to cluster's longevity. Transaction costs among members of local supply chain are reduced as explained above, helping the competitiveness of local supply chain and its survival.

The secondary data show that the number of people employed locally in the production of footwear increased over the years (MTE/RAIS-CAGED). Brito, Brito, Porto & Szilagyi (2010) found that the local agglomeration of members of a supply chain can promote employment growth in the long term as it happens in Jaú.

6. REFERENCES

Asheim, B. T. (2000). Industrial Districts: the contributions of Marshall and beyond. In: Clark, G., Feldman, M., & Gertler, M. S. (Eds.) *The Oxford Handbook of economic geography*. New York: Oxford University Press.

Becattini, G. (1999). Os distritos industriais na Itália. In: Urani, A., Cocco, G., Galvão, A. P. (Eds.) *Empresários e empregos nos novos territórios produtivos*: O caso da terceira Itália. Rio de Janeiro: DP&A Editora.

Brito, E. P., Brito, L. A. L., Porto, E. C., & Szilagyi, M. E. (2010). A relação entre aglomeração produtiva e crescimento: a aplicação de um modelo multinível ao setor industrial paulista. Revista de Administração Contemporânea, 14(4), 615-632.

Cassiolato, J. E., & LASTRES, H. M. (2003). O Foco em Arranjos Produtivos e Inovativos Locais de Micro e Pequenas Empresas. In: Lastres, M. H., Cassiolato, J. E., & Maciel, M. L. (Eds.). *Pequena empresa*: Cooperação e desenvolvimento local. Rio de Janeiro: Relume Dumará.

CETESB. (2014). *Companhia de Tecnologia de Saneamento Ambiental*. Contexto econômico, social e ambiental da Indústria de couros, peles, assemelhados e calçados no Estado de São Paulo. Retrieved from: http://www.cetesb.sp.gov.br/tecnologia-ambiental/cas-em-atividade/44-camara-ambiental-da-industria-de-couros--peles--assemelhados-e-calcados/.

Couromoda (2014). Retrieved from: http://www.couromoda.com/feira/pg/informacoes-gerais/.

Coase, R.H. (1937). The Nature of the Firm. In: Williamson, O. E., & Masten, S. E. (Eds.) *The Economics of Transaction Costs*. Massachusetts: Edward Elgar Publishing, Inc. 1999. pp. 3-22.

Côrrea, A. R. (2001). O complexo coureiro-calçadista brasileiro. *BNDES Setorial*. Retrieved from: http://www.bndes.gov.br/SiteBNDES/export/sites/default/bndes_pt/Galerias/Arquivos/conhecimento/bnset/set1404.pdf.

Dyer, J. H. (1997). Effective Interfirm Collaboration: How Firms Minimize Transaction Costs and Maximize Transaction Value. *Strategic Management Journal*, 18(7), 535-556.

Farina, E. M. M. Q., Saes, M. S. M., & de Azevedo, P. F. (1997). *Competitividade: mercado, estado e organizações*. Editora Singular.

Godoy, A. S. (2006). Estudo de caso qualitativo. In: Godoi, C. K., Bandeira-de-Mello, R., & Silva, A. B. da. (Eds.). *Pesquisa qualitativa em estudos organizacionais*: Paradigmas, estratégias e métodos. 1 ed. São Paulo: Saraiva.

Guidolin, S. M., Costa, A. C. R. da & Rocha, E. R. P. da (2010). Indústria calçadista e estratégias de fortalecimento da competitividade. *BNDES Setorial*. Retrieved from

http://www.bndes.gov.br/SiteBNDES/export/sites/default/bndes_pt/Galerias/Arquivos/conhecimento/bnset/set3104.pdf

Kreps, D. M. (1990). A Course in Microeconomic Theory. New York: Harvester Whatsheaf.

Marshall, A. Princípios de Economia (1891). In: Coleção - *Os Economistas*. São Paulo: Nova Cultural, 1985.

Merriam, S. B. (2002). *Qualitative research in practice*: Examples for discussion and analysis. San Francisco: Jossey Bass.

North, D. C. (1990). *Institutions, Institutional Change and Economic Performance*. Nova York: Cambridge University Press.

Nosso São Paulo. (2014). Cidade de Jaú. Retrieved from: http://www.nossosaopaulo.com.br/Reg_05/Reg05_Jau.htm

Oiko, O. T. (2007). *Desenvolvimento de um sistema de informação para benchmarking e sua aplicação em arranjos produtivos locais*. Dissertação (Mestrado em Engenharia da Produção). Escola de Engenharia de São Carlos. São Carlos.

Oliveira, A. M. R., & Garcia, L.B.R. (2001). O pólo calçadista de Jaú: Suas implicações socioeconômicas e espaciais. In: Gerardi, L. H. de O., & Mendes, I. A. (Eds.). *Teoria, Técnicas, Espaços e Atividades*: Temas de Geografia Contemporânea. Rio Claro: Programa de Pós-Graduação em Geografia UNESP/Associação de Geografia Teorética AGETEO.

Perry, M. (1999). *Small firms and Network Economies*. New York: Routldege.

Porter, M. E. (1991). Towards a dynamic theory of strategy. Strategic Management Journal, 12(S2), 95-117.

Prefeitura Municipal de Jaú. Retrieved from: http://www.jau.sp.gov.br/. Acesso em: Novembro de 2014.

APL de Calçados de Jaú. (2007). *Plano de Desenvolvimento do Arranjo Produtivo Local*. Retrieved from: http://www.desenvolvimento.gov.br/arquivos/dwnl_1248288185.pdf.

Puga, F. P. (2003). *Alternativas de apoio a MPMEs localizadas em arranjos produtivos locais*. Texto para discussão No. 99. Rio de Janeiro: BNDES. Retrieved from: www.bndes.gov.br/.

Robertson, P. L., & Langlois, R. N. (1995). Innovation, networks, and vertical integration. *Research policy, 24*(4), 543-562.

SEBRAE (Serviço de apoio às micro e pequenas empresas de São Paulo). Subsídios para a identificação de clusters no Brasil: atividades da indústria - Relatório de Pesquisa. São Paulo: Sebrae, 2002.

SENAI JAÚ. (2014). Serviço Nacional de Aprendizagem Industrial – Jaú. Retrieved from: http://www.sp.senai.br/home/Telas/enderecos/lista_enderecos.asp?ids=790.

Sengenbreger, W., & Pike, F. (1999). Distritos industriais e recuperação econômica local: questões de pesquisa e de política. In: Urani, A., Cocco, G., & Galvão, A. P. (Eds.) *Empresários e empregos nos novos territórios produtivos*: O caso da terceira Itália. Rio de Janeiro: DP&A Editora.

Suzigan, W., Furtado, J., Garcia, R., & Sampaio, S. (2003). Sistemas locais de produção: mapeamento, tipologia e sugestões de políticas. In: *XXXI Encontro Nacional de Economia* – ANPEC 2003, Porto Seguro. Anais... Porto Seguro, 2003.

Williamson, O. E. (1985). *The Economic Institutions of Capitalism*: Firms, Markets, Relational Contracting. Nova York: The Free Press.

Williamson, O. E. (1996). *The Mechanisms of Governance*. Nova York: Oxford University Press.

Zylbersztajn, D. (1995). *Estruturas de governança e coordenação do agribusiness*: Uma aplicação da Nova Economia das Instituições. 1995. 238p. Tese (Livre-Docência) – Departamento de Administração, Faculdade de Economia, Administração e Contabilidade, Universidade de São Paulo, São Paulo, 1995.

Author's Biography:

Eliane Pereira Zamith Brito: Has more than 15 years of management experience in consumer goods companies. She holds a PhD from Manchester Business School, England, and is a professor at EAESP – Fundação Getulio Vargas, Brazil. Her main research interests are cooperation, trust and reputation in relationships.

Lilian Soares Pereira Carvalho: Visiting researcher at JMSB - Concordia University under Dr. Gad Saad's supervision. Bachelor's at Administração de Empresas from Universidade Presbiteriana Mackenzie (2001) and master at Business Administration from Universidade Presbiteriana Mackenzie (2007). PhD candidate at Fundação Getulio Vargas (FGV-EAESP) since 2012. Research interests: evolutionary psychology, conspicuous consumption.

11

Drivers for implementing environmental requirements – An international explorative study in manufacturing

Anna Sannö

Department of Product Realisation, Mälardalen University, Sweden

anna.sanno@mdh.se

Mats Deleryd

Department of Product Realisation, Mälardalen University, Sweden

mats.deleryd@mdh.se

Anders Fundin

Department of Product Realisation, Mälardalen University, Sweden

anders.fundin@mdh.se

ABSTRACT: Manufacturing companies need to respond to a sustainable development in view of the limitations of planet Earth. This paper explores driving forces for environmentally driven change by gathering interview data from 27 manufacturing plants in Europe and America. A proposed model connects external change triggers with required change actions within organizations.

Keywords: Drivers, Sustainable Operations, Change Management

1. INTRODUCTION

Manufacturing companies must become better at managing the needs coming from environmental requirements in order to establish a sustainable world for future generations. Environmental issues such as energy, natural resources, pollution and waste offer both competitive opportunities and constraints, and are changing the competitive landscape in many industries. For the companies, this will require a responsiveness to external events that leads to the implementation of, and adaptation to, the new environmental requirements. The companies need to respond both by eco-efficiency in business models, in performance measurements and eco-efficiency in products and production (EU 2011). This paper responds to changes towards eco-efficiency in production. Different factors that trigger the change have been explored by authors within the field (Florida et al. 2001; Luken and Van Rompaey 2008; Mittal and Sangwan 2014; Post and Altma 1994). Considering this literature, but also theories from the change management field, this research finds its foundation in the change management framework of Oakland and Tanner (2007). The framework presents a view of relating the external event to the internal need for change in the organization. How the internal need is translated from the external event impacts the process and the final outcome of the change (Oakland and Tanner 2007).

The paper contributes to operation management by exploring changes triggered by environmental requirements in production. While previous research has mainly focused on the external pressures on the organizations, this paper focuses on the sets of drivers that create an implementation process in the production organization. An international explorative multiple case study has been performed with the objective to respond to the following questions "What triggers changes based on environmental requirements?" as well as "What are the drivers for implementing them?".

For operations managers in practice the study should provide support by identifying the different driving forces that can be used as means of creating motivation for internal change projects. The research conducted forms a part of an overall research project aiming to facilitate the implementation of environmental requirements in manufacturing.

2. ENVIRONMENTAL NEEDS AND IMPACT

There is a long-seen need for an environmental, economic and socially sustainable society – a society meeting the present needs without compromising the ability of future generations to meet their own needs (World Commission on Environment and Development 1987). Several efforts have been made both in industry and research (Angell and Klassen 1999; Nunes and Bennett 2010). Nevertheless, new types of products, operations and organization models will be needed to comply with the new constraints and the new objectives of sustainable manufacturing as sustainability itself is dynamic *"a certain situation valid at a certain time can change because of external factors"* (Garetti and Taisch 2011). While the literature in the field has considered drivers from cost advantages, market awareness, life cycle implications and lean and quality integration (Sannö et al. 2014), the future need for change will be derived from earth's capacity and resource depletion (Clift 2005; Perdan 2011).

Post and Altma (1994) provide a view of three drivers for environmentalism: compliance-based, market driven as well as value driven. The value-driven driver provides the understanding that consumers are willing to act on their environmental values. Aligned with this, Luken and Van Rompaey (2008) conclude that the drivers for environmentally sound technology adoption are dependent on subsector, country variations but also the closeness of end-customers. This means that an appropriate strategy to drive environmental change must do more than rely on a traditional regulatory approach; it must also leverage market and community pressures. The community pressures are important for the internal organizations too; Khanna and Anton (2002) find that differences in the environmental practices adopted depending on the incentives were created to meet regulatory threats or to see market opportunities. A third factor, apart from institutional pressure and the ability for organizations to adopt environmental management practices, are the organizational characteristics (Delmas and Toffel 2004). Bey et al. (2013) are also considering the sustaining drivers that are important to keep an implementation going.

The external event and the need for change in the operations form a part of the context of the implementation project (Oakland and Tanner 2007). Jacobsen and Thorsvik (2013) identify three levels influencing the organization. Level 1 is the closest domain including customers, partners and competitors as well as the laws and regulations that only apply to the type of business organization involved. Level 2 comprises the national circumstances consisting of the general political conditions, economic and cultural conditions as well as laws and regulations that apply to all within the nation's borders. Level 3 represents the international and global conditions that are affecting the organization more indirectly but significantly, such as international economic agreements, political events in other countries, climate change and technology. The effect from this level is hard to predict but it also depends on the different pressures and how strongly the organization is affected by the technological and institutional environment. The organization responds to the outside world into three bases for institutional pressure –regulative (legislation), normative (values and norms of society) or cognitive (obvious ideas in the same industry) (Jacobsen and Thorsvik 2013). Organizations that are proactive anticipate changes in the world and are able to act before the pressure to change becomes immediate. However, when acting proactively, even if first move advantages provide competitive opportunities, it is difficult for change agents to create a perceived pressure. That means managing proactive changes is often met with resis-

tance because the members of the organization simply question whether it is necessary (Jacobsen and Thorsvik 2013). Hence, one of the important steps for a successful project is to *"create a sense of urgency"* (Kotter 1996).

This *"sense of urgency"* can be created by legislation (Bey, Hauschild, and McAloone 2013; Luken and Van Rompaey 2008; Mittal and Sangwan 2014). However, Gattiker and Carter (2010) conclude that *"regulation may be a way to force organizations to implement various measures but regulation alone is not sufficient when it comes to gaining buy-in at the level of an individual actor within an organization"*.

To summarize, see Figure 1, one can see that there are multiple ways of creating external and internal drivers for change triggered by environmental requirements. Scholars have explored the external factors or triggers as well as the organizational need for change. Less is done to connect these two driving forces. The external events that trigger the change as well as the related internal need for change are further explored in the conducted case study.

Figure 1 - The environmental requirements created by external events, have different triggers that impact the organization. Within the organization a need for change has to be created in order to create commitment and advancements in the change process.

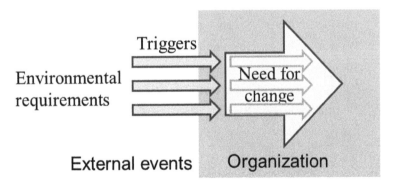

Research Methodology –An industrial case study

A multiple, explorative case study has been conducted in order to create a better understanding of the external events that trigger environmentally driven change. The study also aims for a better understanding of the related need for change that is created in the organization. The methodology was chosen on the grounds of an explorative purpose; the re-

searcher has no control over the events and has a focus on a contemporary phenomenon within a real-life context (Yin 2009). The unit of analysis is drivers for change in the initiation of a change process. The context is change processes within organizations at production plants, with a special interest in changes triggered by environmental requirements.

3. DESCRIPTION OF RESEARCH APPROACH

The selection of sites was made to capture the purpose of the study with a focus on heavy vehicles in the manufacturing industry. The study included 26 plants and one head office, which in total employs about 43,000 people. At the plants 34 people were interviewed during the period from 24 July 2013 to 26 June 2014. The criteria for selecting the cases were large, multinational companies. These companies have production and suppliers available worldwide. Company 1 manufactures vehicles; different business units as well as international production plants were surveyed within the company. As the driver is found to be dependent on the subsector (Luken and Van Rompaey 2008) two companies (7, 8) were chosen having a production of the main chemical processes which create a direct environmental impact in terms of chemicals, waste and emissions. In order to create an understanding of the environmental impacts in the value chain, two companies (6, 9) were selected. Table 1 displays the companies included in the multiple case study.

Table 1 - Overview of the companies included in the multiple case study

Case Company	Subsector	Country (number of plants)	Main Processes	Number of respondents
1	Heavy vehicle	US (1), Brazil (1), Europe (3), Sweden (4)	Assembly, drive train, cab	12
	Trucks & Buses	US (1), Brazil (1), Europe (1), Sweden (4)	Assembly, drive train, cab	9
2	Heavy vehicle	Sweden (1)	Assembly, Component	1
3	Car	Sweden (4)	Assembly, drive train, component, cab	4
4	Trucks	Sweden (1)	Drive-train	1
5	Train	Sweden (1)	Drive-train	2
6	Supplier	Sweden (1)	Component	1
7	Pulp and Paper	Sweden (1)	Chemistry based	2
8	Pharmaceutical	Sweden (1)	Chemistry based	1
9	Retail company	Sweden (1)	No production	1

The interview study has included semi-structured interviews conducted by one of the authors. This method for data collection was chosen on the basis of and developed by principles of Lantz (2013). The key respondents are the environmental coordinator or expert at each production plant. The interview material was first divided into three parts; a background to form an understanding of environmental work, a change project description as well as the future outlook. The interviews lasted for approximately one hour. The interviews have been recorded and transcribed. The use of multiple sources of evidence is used to increase the construct validity according to Yin (2009). Notes, presentations made as well as observations from participation in projects in Company 1 are included in the study in order to triangulate the findings (Yin 2009). The data analysis, within case and cross case analyses of the interviews have been done by the first author and a research colleague by a data analyzing process consisting of pre-coding, categorization and explanation building (Blessing and Chakrabarti 2009; Yin 2009). The results below present a summary of all the responses, and the cross case findings are highlighted in the text. The results and description of each category have been validated by the respondents and also discussed with senior researchers.

4. Empirical findings and discussion

Within the organizations, several parallel projects are ongoing such as energy reduction, waste reduction, change of process technology, implementing control systems as well as having projects to create cultural change. For twenty-six of the respondents the future main focus is predicted to be energy ef-

ficiency. With continuously increasing demands on environmental issues, attention is directed towards maintaining productivity and reducing the cost, global warming and carbon dioxide emissions, waste, chemicals, water, resources and, as some of them state, *"everything"*.

4.1 The external events that will trigger change

With regard to the future events, the respondents respond to what will trigger the changes in the future from an external perspective, see Table 2.

Table 2 - The external event that is predicted to trigger future changes where regulatory, legal requirements are predicted to be the main external driver for change. The number of responses for each driver is highlighted in the bar. Multiple answers are possible.

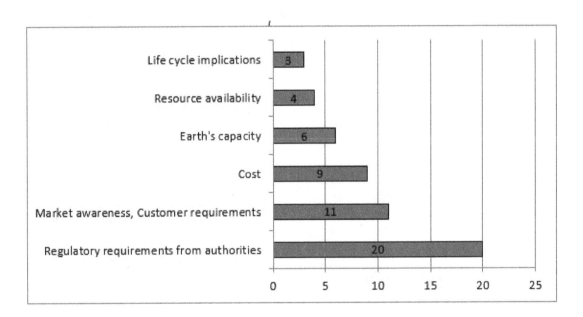

The regulatory requirements from authorities constitute the dominating answer. These requirements are dependent on national and regional legislation. This trigger is also stated to be personal dependent; it is considered to be important to keep good relationships with the personnel at the authorities.

There is a distinction between the different organizations and closeness to the end-customer. For the companies whose products, such as automobiles, pulp and paper to consumer products, the triggers from customer requirements are mentioned as highly important but also the supplier of automotive components.

Cost is also important from two aspects: the respondents mention avoiding penalties but also finding cost advantages. By being recognized as taking en-

vironmental responsibility, the companies maintain competitiveness.

In contrast, the interviewees that mention the earth's capacity and limitations of resources stress this as the driving force that will arise primarily through public opinion. As stated by one of the respondents in the automotive industry, *"What will drive the changes are deficiencies, lack of resources, combined with serious environmental impacts that generate a strong public opinion"*.

That the life cycle implications will drive change is mentioned only in three cases. It is related to having control over the hazardous waste but also the eco-design point of view. Control over the supply chain is of great importance in order to avoid hazardous components in the end-product.

4.1 The need for change

The respondents also respond to what is creating a drive for change within their production plant. Table 3 displays what is creating the need for change within the production plant.

Table 3 - The drivers for creating a need for change within the production plants. Multiple answers are possible.

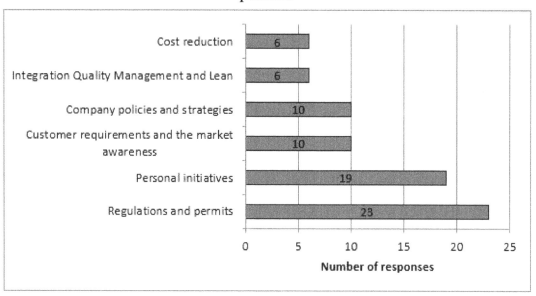

The main factor for creating a need for change internally is the rules and laws, which are connected to the triggers for change being regulatory and to the requirements from authorities. One of the respondents explains this as: *"The simplest is based on legislation, there is no doubt, and we must do it. Not to say the best way to do it, but if there are laws behind they listen to you. The best way is that the act by thinking people, people feel that we need it. But today, legislation is the easiest way to get the people in production to listen to you. "*

Strongly related to regulations and permits, in eleven of the cases, regulation is mentioned as being supported by personal initiative within the organization and the ability to manage the work to meet the statutory requirement. The environmental legislation and controls are continuously updated and the organizations must have procedures to bring in, spread and consider the consequences of the new requirements.

What will drive changes triggered by customer requirements will typically be both cost-related reasons and promotional reasons for increased sales. *"If you are not environmentally aware you will not be able to sell".*

Company policies and strategies of these companies are created on a corporate level and then introduced at the plant level. Here it seems as if there is a difference between the international plants; in Brazil and Germany the company strategy requirements are creating the drive rather than the local laws. *"What is coming from the higher management is the reason why we are doing it. Yes it is. But not laws, we are already meeting the expectations for legislation here."*

Several of the plants have created the internal drive for change by identifying their own environmental aspects by principles within the lean production system. To promote environmental issues with quality - and with lean concepts - helps to bring up environmental issues on the agenda and that the facility may be a systematic approach to work through the process.

Cost reduction is considered to be a force for change to give attention to the change within the internal organization, as for projects related to reducing energy consumption and the amount of waste. These changes have an advantage in that they can get into the strategic plan with clear monitoring and measurable KPIs, and then it creates the attention and drive from management.

Identified differences between the nine case companies

- The differences and similarities notified in the empirical study are presented in the bullets below:

- The subsector is important in relation to the end-customer. It is then competitiveness in the form of public opinion and ultimately the individual consumer choice and market forces that drive the change within the organization. For example, the drive from customer requirements is less mentioned as a driver for change in the production plant within the heavy vehicle sector.

- In Brazil and Germany, where local legislation is providing less drive compared to countries such as Sweden and France, the company policies and strategies are considered as an important driver for change.

- On a strategic level, the company policies and strategies are managed differently on the production plant level. There is more difficulty in driving change if the environmental policy or environmental strategy is introduced to the environmental coordinator. These persons must first gain management support. If strategies are directed to the plant management, the environmental coordinator acts as a support.

- The legal requirements are supported by personal initiatives and commitment.

- In the end, as stated by two of the respondents, the momentum will be the company's development. The development of the production system will be based on the product development towards less environmental impact, but also on the business and the results. The recession in 2008 is remembered as *"becoming a lesser focus on these issues apart from what is licensed and legally controlled"*.

- Multiple ways to drive change are considered by some of the respondents. *"You cannot say that it is the environment that drives, you cannot say that it is money that drives. You cannot say that it is quality that drives. Everything is connected. But when it comes to for example Best Available Technology, BAT, requirements then its demands come from the environment. "*

- The facilities located at a smaller place gain from being located where employees both work and live. The culture for environmental commitment and a drive to manage environmental issues is stronger if the facility is located near summer cottages or the local school.

5. DISCUSSION AND CONCLUSIONS

As a result from the study, it seems that for the multinational companies in the study it is the regulatory events that bring the most pressure on the production plants but also create the need for change internally. This is aligned with the findings in literature (Bey, Hauschild, and McAloone 2013; Luken and Van Rompaey 2008). The result from the case study also clarifies that regulation is linked with personal initiatives, which is aligned with the conclusion by Gattiker and Carter (2010). However, the increased general awareness of the climate change and resource depletion can create personal engagement and public opinion both outside and within the company. The model developed based on the literature and the empirical findings is presented in Figure 2. The model describes the different triggers for change, identified in the literature.

Figure 2 - Visualisation of the production plants' responsiveness for future environmental requirements. From the left the four bases for pressure are shown as technological pressure, regulative (legislation), normative (values and norms of society) or cognitive (obvious ideas in the same industry). The pressures from regulation, customers and the need for considering the life cycle implications are dependent on the subsector. The geography surrounding the plant influences the organizational focus for different environmental issues.

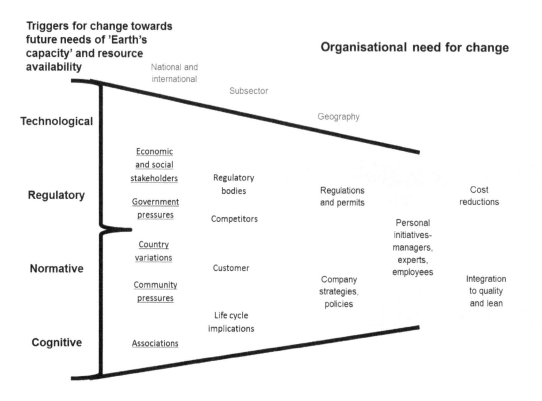

Four pressures will influence the organization directly or indirectly. The pressures will come from different levels, national, international as well as the influence of the environmental requirements from the different subsectors. The geography influences the culture of the organization according to the empirical findings. The model also presents the organizational need for change, where the drive for change is created by regulations and permits, company strategies and policies, cost reductions, integration with quality and lean systems. However, the main driver for change as well as to regulations is seen to be personal initiatives at different levels of the company.

A concern of validity is seen by only including key informants as being the environmental professionals at each site. The perception of this professional group might differ from the perception of plant managers and other stakeholders within the large plants (Luken and Van Rompaey 2008). On the other hand, these environmental professionals initiate a great number of the environmental projects in the organizations studied. By including different subsectors in the analysis, the findings are considered to be generalizable to several subsectors.

6. CONTRIBUTION AND FUTURE RESEARCH

This paper presents different triggers based on environmental requirements and related drivers for implementing them. A model, combining theory and a multiple case study is developed. The contribution to academia is a better understanding of the sets of drivers that create an implementation of change, triggered by environmental requirements. The contribution to practice is a model that could support operational managers in their corporate foresight as well as better understand what drivers can create an internal need for change. It is recommended to conduct further empirical research to validate the model by empirical studies as well as research on how to apply theories from organizational change management within this, in the authors' opinion, important field of operations management.

7. ACKNOWLEDGEMENTS

This research project has been funded by the Knowledge Foundation within the framework of the IN-NOFACTURE Research School and the participating companies, and Mälardalen University, Sweden. The research project is also a part of the initiative for Excellence in Production Research (XPRES) which is a joint project between Mälardalen University, the Royal Institute of Technology, and Swerea. XPRES is one of two governmentally funded Swedish strategic initiatives for research excellence in Production Engineering. The people that contributed with their knowledge are gratefully thanked for making these experiences available.

8. REFERENCES

Angell, L.C., and R.D. Klassen. 1999. "Integrating Environmental Issues into the Mainstream: an Agenda for Research in Operations Management." *Journal of Operations Management* 17 (5):575-598.

Bey, N, M.Z Hauschild, and T.C McAloone. 2013. "Drivers and barriers for implementation of environmental strategies in manufacturing companies." *CIRP Annals -Manufacturing Technology 62* 62 (1):43-46.

Blessing, L, and A Chakrabarti. 2009. *DRM, a Design Research Methodology*:London/New York, Springer.

Clift, R. 2005. "Sustainable development and its implications for chemical engineering." *Chemical Engineering Science* 61 (13):4179-4187.

Delmas, Magali, and Michael W Toffel. 2004. "Stakeholders and environmental management practices: an institutional framework." *Business strategy and the environment* 13 (4):209-222.

EU. 2011. "Eco-innovation action plan." Available at http:// ec.europa.eu/environment/ecoap/about-action-plan/objectives-methodology/index_en.htm, Accessed. 2014. Oct.

Florida, Richard, Mark Atlas, and Matt Cline. 2001. "What Makes Companies Green? Organizational and Geographic Factors in the Adoption of Environmental Practices*." *Economic Geography* 77 (3):209-224.

Garetti, M., and M. Taisch. 2011. "Sustainable manufacturing: trends and research challenges." *Production Planning and Control: The management of operations* 23 (2-3):83-104.

Gattiker, T.F., and C.R. Carter. 2010. "Understanding project champions' ability to gain intra-organizational commitment for environmental projects." *Journal of Operations Management* 28 (1):72-85.

Jacobsen, D-I, and J Thorsvik. 2013. *Hur moderna organisationer fungerar*. Bergen, Norway: Studentlitteratur.

Khanna, Madhu, and Wilma Rose Q Anton. 2002. "What is driving corporate environmentalism: Opportunity or threat?" *Corporate Environmental Strategy* 9 (4):409-417.

Kotter, J.P. 1996. *Leading Change*. Boston, MA.: Harvard Business School Press.

Lantz, Annika. 2013. "Intervjumetodik. 3. uppl." *Lund: Studentlitteratur*.

Luken, R., and F Van Rompaey. 2008. "Drivers for and barriers to environmentally sound technology adoption by manufacturing plants in nine developing countries." *Journal of Cleaner Production* 16 (1):67-77.

Mittal, V.K., and K.S. Sangwan. 2014. Prioritizing Drivers for Green Manufacturing: Environmental, Social and Economic Perspectives. Paper presented at the Procedia CIRP 2015, 21st CIRP Conference on Life Cycle Engineering.

Nunes, B., and D. Bennett. 2010. "Green operations initiatives in the automotive industry -An environmental report's analysis and benchmarking study." *Benchmarking: An International Journal* 17 (3):396-420.

Oakland, J.S., and S. J. Tanner. 2007. "Successful Change Management." *Total Quality Management & Business Excellence* 18 (1-2):1-19.

Perdan, S. 2011. "The Concept of Sustainable Development and its Practical Implications." In *Sustainable Development in Practice. Case Studies for Engineers and Scientists, Second Edition*, edited by S Perdan and A Azapagic, 3-25. Chichester, UK: John Wiley & Sons, Ltd.

Post, James E, and Barbara W Altma. 1994. "Managing the environmental change process: barriers and opportunities." *Journal of Organizational Change Management* 7 (4):64-81.

Sannö, A., L. Stålberg, and A. Fundin. 2014. "Managing environmental change - a movement to reactive to proactive behaviour." *Int. J. Productivity and Quality Management* Accepted for publication October 2014.

World Commission on Environment and Development, WCED. 1987. *Our Common Future (Brundtland Report)*: Oxford University Press.

Yin, R.K. 2009. *Case Study Research -Design and Methods*. Edited by Leonard Bickman and D.J. Rig. Vol. 5, *Fourth Edition*. London: SAGE Publications, Inc.

Author's Biography:

Anna Sannö is an industrial PhD student within the research profile of Innovation and Product Realisation at Mälardalen University and she is in parallel employed by Volvo CE. Her research interest lies in the field of sustainable operations management with a focus on managing change in production systems triggered by environmental requirements. Before her PhD studies, she has several years of industrial practice working with paint and surface treatment within the Volvo Group and in paint companies.

Mats Deleryd is Professor in Quality and Organisational Development at Mälardalen University and Senior Vice President "Quality, Safety & Environmental Care" at Volvo Group based out of Gothenburg, Sweden. His research interests are within Quality- Environmental- and Sustainability Management. Mats is chairman of the board of the research school INNOFACTURE. He is also chairman of the board of the Swedish Institute for Quality, SIQ. In 2014 he was also elected academician and member of the International Academy for Quality, IAQ.

Anders Fundin is full Professor in Quality Technology and Management at Mälardalen University. His research interest is related with Quality Management, Operational Development, Lean Production, and Continuous Improvement. His research work has been performed in collaboration with about thirty multinational companies and Anders has more than 15 years of research and development experience within the automotive industry. He has written about 35 papers presented at international conferences or published in international journals such as for example International Journal of Service Industry Management, International Journal of Quality and Reliability Management, Journal of Business and Industrial Marketing and International Journal of Productivity and Quality Management.

12

Is there a Link between Supply Chain Strategies and Firm Performance? Evidence from Large-Scale Manufacturing Firms in Kenya

Richard Bitange Nyaoga
Egerton University
rnyaoga@yahoo.com

Peterson Obara Magutu
University of Nairobi
magutumop@yahoo.com

Josiah Aduda
University of Nairobi
jaduda@uonbi.ac.ke

ABSTRACT: The purpose of this study was to establish the relationship between supply chain strategies and performance of large-scale manufacturing firms in Kenya by addressing three primary gaps in the literature. The research gaps include the research findings and results on the relationship between supply chain strategies and firm performance that have been contradicting and no attempt to clear the contradictions; biased and unbalanced analysis of the different measures of firm performance, and failure to use weighted scores to measure firm performance. Resource-Based View guided this study. A sample of one hundred and thirty-eight (138) firms was drawn using proportionate sampling from a population of six hundred and twenty-seven (627) large-scale manufacturing firms in Kenya. The response rate was seventy-five (75) percent. The correlation analysis and regression analysis models were used to test the hypotheses. The study findings indicate that Supply chain strategies are useful predictors of the firm's performance as supply chain strategies explain 76.7 % of the changes in the firm's performance.

Keywords: Supply chain management, Supply chain strategies, Firm performance

1. INTRODUCTION

The concept of supply chain management (SCM) has been the subject of numerous studies in operational management, purchasing, logistics, and marketing. There are a number of constructs like supply chain strategy and technology that can be used in these diverse research domains as there is lack of empirical evidence in SCM practice (Halley & Beaulieu, 2009; Hult, Ketchen, Cavusgil, & Calantone, 2006; Ketchen & Giunipero, 2004). Owing to lack of consensus on definition and differing views on the concept of SCM, this study was guided by Mentzer et al. (2001). Mentzer et al. (2001) definition that is broad enough and captures the issues of strategy and firm performance. They define supply chain management as:

> "...the systemic, strategic coordination of the traditional business functions and the tactics across these business functions within a particular company and across businesses within the supply chain, for the purposes of improving the long-term performance of the individual companies and the supply chain as a whole (Mentzer et al., 2001, p. 18)".

The Resource Based View (RBV) and Transactional theories have played a very crucial role when conducting research on the strategic perspectives of operations and supply chain management (Burgess, Singh, & Koroglu, 2006). The resource-based view theory has been greatly used in SCM studies in the last twenty years. This theory has to a great extent shaped mastery of operational decisions in the context of SCM (Halley & Beaulieu, 2009; Patterson, Grimm, & Corsi, 2003).

Under the economic pillar of Kenya Vision, 2030, manufacturing is one the five sectors that has been identified to support economic development. In line with the aspirations of Vision 2030, it is expected to be a dominant and aggressive sector to support the national growth, create employment, earn the country foreign exchange and facilitate foreign investment (GoK, 2007). Many large-scale manufacturing subsector companies in Kenya particularly multinational manufacturing firms have migrated their operations to other countries. These firms have relocated, shut down or downsized their operations because they consider Kenya as one of the least yielding countries worldwide. This is due to poor infrastructure, high tariffs and taxes. The local firms have not been able to fill the manufacturing gaps left by the multinationals as the government has done

very little to develop this struggling subsector leading to low international competitiveness (Okoth, 2012; PwCIL, 2010).

Hines (2009) define what the supply chain strategies are, how they work and why firms invest in them as follows:

> "Supply chain strategies require a total systems view of the linkages in the chain that work together efficiently to create customer satisfaction at the end point of delivery to the consumer. As a consequence, costs must be lowered throughout the chain by driving out unnecessary costs and focusing attention on adding value. Throughput efficiency must be increased, bottlenecks removed and performance measurement must focus on total systems efficiency and equitable reward distribution to those in the supply chain adding value. The supply chain system must be responsive to customer requirements."

In essence, research indicates that there are sixteen supply chain strategies in use today. These include: synergistic; project logistics; Nano-chain; information networks; market dominance; value chain; extended; efficient; cash-to-cash cycle; innovation; speed to market; risk-hedging; micro-chain; tie down; none existent; and demand supply chain strategies. There are some benefits, challenges, and relative complexity for each of these sixteen supply chain strategies. This has led to the categorization of the sixteen supply chain strategies into a dichotomy of Long-range and Mid-range supply chain strategies (Gattorna, 2006; Gadde, & Hakansson, 2001). The sixteen-supply chain strategy dichotomy was central in this study about firm performance. This study considered both the direct effect of these long-range and mid-range supply chain strategies on the performance of large-scale manufacturing firms in Kenya.

Performance management is crucial and a paramount practice to the success of any business. Being a wide-ranging topic, one can focus on target/ goal setting, measurement, feedback or reward. The biggest challenge facing firm performance measurement is most scholars limit themselves to their areas of specialization; few academics across these functional boundaries to make reference to the research of other experts outside their functional areas (Neely, Bourne, & Kennerley, 2000). This study measured performance using indicators cutting across all functional areas in relation to firm performance.

Organizations in today's business environment have a big challenge on how to remain competitive in the marketplace through firm performance especially the organization-wide performance (Collins, Worthington, Reyes, & Romero, 2010). Agha, Alrubaiee, & Jamhour (2011), argued that to remain competitive and achieve competitive advantages, managers should increase organizational performance by managing the dimensions of core competence i.e. a shared vision; cooperation and empowerment. Some authors like Keegan, Eiler, & Charles (1989) and Kaplan & Norton (1992) have suggested appropriate firm performance measurement frameworks to the management community. They include the performance measurement matrix and the Balanced Scorecard (BSC). The performance measurement matrix as advanced by Keegan et al. (1989) ranks activities in matrix form, but it does not assign weights. The BSC is a tool that gives a balanced view of how an organization is performing by cascading firm performance perspectives down from the strategic to operational level of business at four levels of: customer service; financial stewardship; internal business processes; and service delivery innovations (Gunasekaran, Patel, & Tirtiroglu, 2001). The BSC has simplified the measurement of firm performance, especially for supply chains where all units share the metrics in the organization and supply chain partners (Kaplan & Norton, 1992). According to Bolo (2011) the concept of firm performance and its measurement has not been extended beyond the firm's inbound operations. This limited visibility of measures tends to exclude SC performance measures. This study explored the balanced approach for firm performance with four perspectives within the context of large-scale manufacturing firms in Kenya.

1.1. Research Problem and Research Focus

According to Cooper & Ellram (1993), SCM is an integrative philosophy to manage the total flow of goods from the supplier to the ultimate user. Varma, Wadhwa, & Deshmukh (2006) considers this definition of SCM as a management philosophy that tries to bring about integration among various functions. Cousins, Lawson, & Squire (2006), definition is comprehensive and critical, they describe the concept as consisting of the flow of raw materials, finished goods, finances and information while aiming to achieve high customer expectations through proper planning on demand forecasts, sales generation, and efficient distribution. The flow should be well coordinated in the form of a network starting with suppliers then to manufacturers, next to distributors

and finally customers. But this definition does not include the elements of uncertainty which require some element of strategic focus. According to Kamaruddin & Udin (2009) there is a high degree of uncertainty and difficulty in managing supply chains, especially where there are multiple relationships and interactions between elements of the firm's existing network.

According to PwCIL (2010) and Okoth (2012), Kenya's large-scale manufacturing subsector has a challenging history in terms of performance and unstructured strategy. This study sought to test contextually the relationship between SC strategies and performance of large-scale manufacturing firms in Kenya. As observed by Burgess et al. (2006) most of the researches done on SCM is on very few industries covering the consumer goods retailing, computer assembling and automobile manufacturing. This study overcame this by covering twelve subsectors of the large-scale manufacturing firms in Kenya.

An expanded approach of sixteen-supply chain strategies dichotomy is in use today, and the future shall see firms competing using their supply chains strategies (Gadde & Hakansson, 2001). Very few studies have attempted to address such an expanded approach to sixteen SC strategies in establishing the relationship between supply chain strategy and firm performance (Gattorna, 2006; Russell & Hoag, 2004). The sixteen-supply chain strategy dichotomy provides an extended approach whose relationship with firm performances are the subject of this study.

Most studies have therefore used a limited number of measures that are not objective enough to establish a link with the concepts studied. Mainly, they have not used the Balanced Scorecard to determine firm performance something the current study sought to use. This was therefore guided by the following research question: What is the relationship between SC strategies and firm performance? The main objective of this study was to establish the relationship between supply chain strategies and performance of large-scale manufacturing firms in Kenya.

1.2. Empirical studies

Teeratansirikool, Siengthai, Badir, & Charoenngam (2013) argued that all competitive strategies positively and significantly enhance firm performance. Khan & Pillania (2008), argued that supplier evaluation, strategic supplier partnership, sourcing flexibility and trust in supply chain members have a significant effect on supply agility and firm's performance. Qrunfleh & Tarafdar (2014) posits that in small firms, efficient Supply Chain integration plays a more critical function for sustainable performance improvement, while, in large firms, the close inter-relationship between the level of SCM practices and competition capability have more significant effect on performance improvement.

Li, Ragu-Nathan, Ragu-Nathan, & Subba Rao (2006) in their study they conceptualized on five dimensions of SCM practices (customer relationship, quality of information sharing, strategic supplier partnership, level of information sharing, and postponement) and tested the relationships between competitive advantage, SCM practices, and organizational performance. Their results indicate that higher levels of SCM practices can lead to improved competitive advantage and enhanced firm performance.

Ou, Liu, Hung, & Yen (2010) found out that external customer-firm-supplier relation management positively influence firm internal contextual factors that in turn have positive effects on firm performance. Their finding indicates that successful implementation of SCM practices directly improves operational performance, and also indirectly enhances customer satisfaction and financial performance of the organization.

Golicic & Smith (2013) concluded that the relationship between environmental supply chain practices and operational-based, market-based and accounting-based forms of firm performance are positive and significant, giving support for the argument that sustainable supply chain management leads to increased firm performance. Zolait, Ibrahim, Chandran, Pandiyan, & Sundram (2010) in their study argued that information flow, financial flow, and physical flow were statistically significant to firm performance. Lee, Kim, & Choi (2012) argued that there are a significant indirect relationship between Green Supply Chain Management (GSCM) practice implementation and firm performance through mediating variables of operational efficiency and relational efficiency.

1.3. Conceptual Model and Hypothesis

The conceptual model in figure 1 below is in support for the arguments raised from literature review that the SC strategies that consist of Mid-range SC strategies and Long-range SC strategies have a relationship with firm performance outcome of large-scale manufacturing firms in Kenya. Figure 1 below is emphasizing the interconnection between the SC strategies and firm performance in one comprehensive framework intended to assist the researcher in developing a clear understanding of the linkages between the two variables.

Figure 1: Conceptual Model

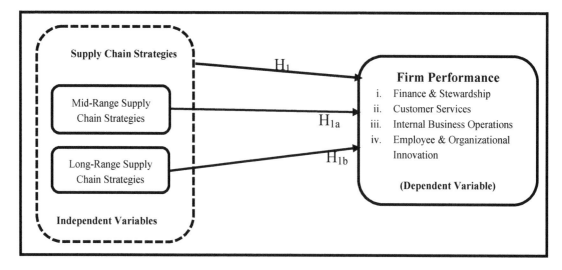

Source: Author, 2014

Is there a Link between Supply Chain Strategies and Firm Performance? Evidence from Large-Scale Manufacturing...

187

Based on the study objective, this study examined the supply chain strategies that consist of Mid-range SC strategies and Long-range SC strategies and their relationship with firm performance. Hence, the following hypotheses were tested:

H1: Supply chain strategies are positively related to firm performance

Given that the sixteen-supply chain strategy dichotomy (Mid-range SC strategies and Long-range SC strategies) was used as independent variables about firm performance, the following two sub-hypotheses were derived from the hypothesis two.

H1a: Mid-range SC strategies are positively related to firm performance

H1b: Long-range SC strategies are positively related to firm performance

2. METHODOLOGY

2.1. General Background of Research

The positivistic philosophy was preferred to guide this study since it combines static and a priori approaches. The positivistic paradigm often requires a test of a model using questionnaires constructed without input from the respondents as it was the case for this study. Moreover, this research comprised of predefined (a priori) relationships that required primarily theory testing as all the hypotheses are stated with predictive rigor for acceptance aimed at making positivistic conclusions.

2.2. Research Design

This study adopted a cross-sectional survey and descriptive design. The design was appropriate because it is useful in establishing the nature of existing situation and current conditions and also in analyzing such situations and conditions. Johnson, Scholes, & Whittington (2007) did a similar study in USA and used the same methodology and variables. Fawcett, Ogden, Magnan, & Cooper (2006) used strategy implementation as the independent variable and performance as the dependent variable using a triangulation methodology consisting of literature review, survey, and case studies. Given this approach, a cross-sectional survey method was used to obtain the empirical data to determine the linkages between variables.

2.3. Population of the Study

The target population was all large-scale manufacturing firms in Kenya. The unit of analysis was the large scale manufacturing firm. In Kenya, according to the KAM directory (2010/2011) large scale enterprises have more than 100 workers, medium enterprises have from 51 to 100 workers, small enterprises have from 11 to 50 workers, and micro-enterprises are those with 10 or fewer workers. There are 2,000 manufacturing companies in Kenya, from which the target population is 627 large-scale manufacturing firms. Although the categorizations of manufacturing firms according to size has been based on the number of employees, the type and level of technology used, size of capital investment and capacity utilization can be used to justify the choice of large-scale manufacturing firms. The main reason for this choice is that these firms are likely to exhibit an elaborate SCM philosophy, exhibit high activity levels, have enough resource to be employed in supply chain strategy implementation, make use of supply chain strategies and SCT in SCM. The number of employees is a good indicator of size because being profit making; employees can be taken as a proxy for supply chain performance, profits, technology utilization and firm performance. Large-scale manufacturing firms that make more than two-thirds of the industrial coverage is considered as the strength of this research since prior studies had ignored sector-specific supply chain variables on firm performance.

2.4. Sample of Research

The appropriate sample size for a population-based survey was determined largely by three factors (Kate, 2006): (i) the estimated percentage prevalence of the population of interest – 10% (ii) the desired level of confidence and (iii) the acceptable margin of error.

The sample size required can be calculated according to the following Kate (2006) formula

$$\mathbf{n} = \frac{t^2 \times p(1-p)}{m^2}$$

Where:

n = required a sample size, t = confidence level at 95% (standard value of 1.96), p = estimated percentage prevalence of the population of interest – 10%, m = margin of error at 5% (standard value of 0.05)

Therefore, the sample size (n) for this study can be computed as follows:

$$n = \frac{1.96^2 \times .1(1-.1)}{.05^2}$$

$$n = \frac{3.8416 \times .09}{.0025}$$

$$n = \frac{.3457}{.0025}$$

$$n = 138.30 \sim \mathbf{138}$$

One hundred and thirty-eight (138) large scale manufacturing firms were sampled and contacted to participate in the study. Then the large-scale manufacturing firms (sample) were stratified into twelve key sectors/strata as shown in table 1 based on the KAM directory of 2010/2011.

Table 1: Sampling Strata

Large-Scale Manufacturing Sectors/Strata	Strata Popn N	Proportionate Sampling P_n=N/Total Popn *Sample
Building, Construction, and Mining	15	3
Food, Beverages, and Tobacco	154	33
Chemical and Allied	71	16
Energy, Electrical and Electronics	43	10
Plastics and Rubber	66	14
Textile and Apparels	68	15
Timber, Wood Products, and Furniture	26	6
Pharmaceutical and Medical Equipment	32	7
Metal and Allied	62	14
Leather Products and Footwear	8	2
Motor Vehicle Assembly and Accessories	22	5
Paper and Paperboard	60	13
Total	**627**	**138**

Source: Researcher, (2014)

Proportionate sampling was done to pick the required number of respondents from the twelve (12) strata. This gave every firm an opportunity to participate in the study.

2.5. Instrument and Procedures

Data for this study was collected from both primary and secondary sources that are meant to reinforce each other (Stiles, 2003). Primary data entailed responses on all the study variables: supply chain strategies and firm performance. Secondary data, particularly five-year historical data on firm performance data was sourced from company annual reports, pamphlets, office manuals circulars, policy papers, corporate /business plans as well as survey reports from Kenya Association of Manufacturers and Kenya Central Bureau of Statistics for the years 2006 - 2010. This is because the normal planning cycle at the strategic level is five years.

The questionnaire and data forms were the principal tools for collecting primary data and secondary data respectively. One respondent, either the Operations Manager or Supply Chain Management Manager or procurement manager from each firm, was selected to participate in the study. Wilson & Lilien (1992) showed that single informants are most appropriate in non-new task decisions. Based on this, the criterion for choice of a respondent in each firm is that one should be experienced or knowledgeable about the supply chain management, operations management decisions and activities of the firm at the time of the survey. The researcher administered the questionnaires personally in order to enhance the response rate and quality of data collected as supported by Bhagwat & Sharma (2007) using the official request.

2.6. Data Analysis

The positivistic approach that advocates for hypotheses testing using quantitative techniques to research guided data analysis (Stiles, 2003).Thus, information required for testing the study hypotheses was generated using quantitative data analytical techniques. Consequently, data analysis followed Umma (2006) four step process for data analysis: "getting data ready for analysis; getting a feel for the data; testing the goodness for the data; and testing the hypotheses".

The researcher used descriptive statistics for Likert scale variables in the questionnaire. The measures of dispersion were used in order to explore the underlying features in the data on large scale manufacturing firms in Nairobi, Kenya. Descriptive statistics covered all response variables as well as the demographic characteristics of respondents. Descriptive statistics provides the basic features of the data collected on the variables and provide the impetus for conducting further analyzes on the data (Ezirim & Nwokah, 2009; Mugenda & Mugenda, 2003).

A correlation analysis was done to establish the relationships among the study variables. The correlation analysis was computed to describe the relationships that exist among key variables of the study and/or use the known correlation to determine the outcome from one variable to another. A multiple linear regression model was adopted to study the linear relationships among the various study variables. A multiple linear regression analysis is a multivariate statistical technique used to estimate the model parameters and determine the effect of individual independent variables (IVs) on the dependent variable (DV).

Firm performance

$$(Y) = b0 + b1X1 + b2X2 + b3X3 + b4X4 \ldots + bpXp + ei \quad (1)$$

Where;

Y is the dependent variable (Firm Performance) and is a linear function of X1, X2, X3, X4…Xi plus ei .

Y Firm Performance Index (FPI) was computed as an average of the five year's Annual Firm Performance Composite.

b0 is the regression constant or intercept, b1-p are the regression coefficients or change induced in Y by each X, ei is a random variable, error term that accounts for the variability in Y that cannot be explained by the linear effect of the i predictor variables and X1-p   are independent variables (Long-range and Mid-range supply chain strategies)

3. RESEARCH RESULTS AND FINDINGS

One of the methodological weaknesses of previous studies were small sample sizes and low response rate. This study's response rate of 75% is high compared to previous studies whose average response rate was 65 percent or less. Kidombo (2007) who had studied large private manufacturing firms in Kenya had 64% response rate; Kirchoff (2011) had a very low response rate of 28 percent. According to Tomaskovic-devey et al. (2007) any response rate of about 15.4% is considered as yielding a relatively high response rate considering the demands on the time of top-level executives. All subsectors of the large scale manufacturing sector were well represented in this study, avoiding any chances of bias or misrepresentation.

The majority of the firms (68%) have successfully managed their supply chains while 16% see their supply chains as very successful and somewhat successful. This is an indication that the supply chain department exists in most large-scale manufacturing firms (84) and probably managed by specialists who understood what the items in the questionnaire were testing and the appropriate response that was required. This implies that only those firms that have managed their supply chains have sound strategies that are in place to guide the operations of the firm.

3.1. Firm Performance Index

Weighted scores were applied on the collected data to determine the firm performance index on average for all the firms that participated in this study as shown in Table 2 below.

Manufacturing Technology and Production Management

Table 2: Firm Average Performance Index

DOMAIN			ACHIEVEMENTS					Weighted Performance (WP i) ... (4)					
			2006	2007	2008	2009	2010	2006	2007	2008	2009	2010	
A. Financial & Stewardship													
Pre-tax Profits	Kshs.(m)	10	108.6	132.1	137.2	131.8	158.0	10.86	13.21	13.72549	13.18	15.80718	
Debt –Equity Ratio	%	5	38.3	42.1	48.0	47.2	50.5	1.91	2.10	2.40	2.36	2.525	
Return on Investment	%	5	41.7	45.9	51.1	53.7	57.5	2.08	2.29	2.55	2.68	2.87	
Development Index	%	5	44.7	49.5	55.4	60.1	66.0	2.23	2.47	2.77	3.00	3.30	
Payback on investments	Time	5	8.7	8.2	7.3	6.4	6.9	0.43	0.41	0.36	0.32	0.34	
Weights – Sub Total		30	242.2	277.9	299.8	299.5	339.2	17.54	20.50	21.817	21.56	24.86	
B. Customers Perspective													
Customer satisfaction	%	10	61.3	65.7	70.8	75.8	79.7	6.13	6.57	7.08	7.58	7.97	
Customer price margin	%	6	39.8	41.8	43.4	45.5	46.5	2.39	2.50	2.60	2.73	2.79	
Resolution of customer complaints	%	4	60.0	64.7	69.5	73.4	79.2	2.40	2.59	2.78	2.93	3.17	
Weights – Sub Total		20	161	172	183	194	205	10.9	11.7	12.5	13.3	13.9	
C. Internal Business Operations													
Cost efficiency	%	10	55.9	59.8	64.3	68.5	73.3	5.59	5.98	6.43	6.85	7.33	
Automation	%	8	50.9	56.2	61.7	66.3	72.7	4.07	4.50	4.93	5.30	5.82	
Warranty quality	%	6	55.9	59.6	63.8	68.3	73.3	3.35	3.58	3.83	4.10	4.39	
Safety Measures	%	2	59.1	63.9	68.9	72.0	78.9	1.18	1.27	1.37	1.44	1.57	
Research & Development	%	4	51.5	56.3	61.0	65.1	71.9	2.06	2.25	2.449	2.60	2.87	
Work Environment	%	2	56.3	60.4	64.9	68.5	73.6	1.12	1.20	1.29	1.37	1.47	
Capacity Utilization	%	4	58.2	62.3	67.9	72.3	77.9	2.33	2.49	2.71	2.89	3.11	
ISO Certification (9001:2008)	%	4	44.3	47.6	55.7	59.9	65.5	1.77	1.90	2.22	2.39	2.62	
Weights – Sub Total		40	432	466	508	543	587	21.5	23.2	25.3	27.0	29.2	*Average FPI ——6*
D. Employee and Organization Innovation													
Employee satisfaction	%	5	59.4	64.2	68.1	71.8	75.8	2.97	3.21	3.409	3.59	3.79	
Employee Retention	%	2	75.6	79.8	81.9	85.7	87.3	1.512	1.596	1.638	1.714	1.746	
Employee productivity	%	2	60.5	65.0	69.4	72.0	75.5	1.21	1.30	1.38	1.44	1.51	
Competency Development	%	1	56.6	61.4	65.5	69.6	73.4	0.56	0.61	0.65	0.69	0.73	
Weights – Sub Total		10	252	270	285	299	312	6.3	6.7	7.1	7.4	7.8	
TOTAL/Performance Index		100	Annual Firm Performance......5					56.2	62.1	66.6	69.2	75.8	66.0

Source: Research Data, 2014

From the results in Table 2 above on firm performance, there is specific improvement on the four dimensions of firm performance of financial & stewardship, customers' perspective, internal business operations including those of employee and organization innovation. This is an indication that the firms have improved performance that is balanced touching on all aspects of the firm about its internal and external customers who make up its supply chain. All the four domains were equally affected in 2008/2009 period that might be as a result of the post-election violence in Kenya. Each of the four dimensions of the firm's performance is a relative sector to the total sub weights. For example, in 2006 the firms scored 6.3 out of the possible score of 10% in the employee and organization innovation (x/10).

The above computations were done for each firm to determine their annual firm performance and firm performance index that was used as the dependent variables (Y) in the next section of correlation analysis and subsequently on test of hypotheses.

3.2. The Correlation between Supply Chain Strategies and Firm Performance

Spearman's rank order correlation analyzes the relationships between supply chain strategies (Mid-range and long-range) and firm performance as presented in Table 3 below. From the results in Table 3, there is a strong, and positive relationships are observed between long-range supply chain strategies (r = 0. 690, p< 0.01) and firm performance. These two long-range supply chain strategies are demand supply chain strategy and innovation supply chain strategy. Indeed, innovations and demand are specific the firms' operations and products respectively.

Table 3: Correlation between Supply Chain Strategies and Firm Performance

Supply Chain Strategy	Variables	Spearman's rho Coefficients
Long-range SC Strategy	Innovation SC strategy.	0. 690(**)
Mid-range SC Strategy	No need for SC strategy	0.591(*)
Long-range SC Strategy	Demand SC strategy.	0.545(*)

Source: Research Data, 2014

** Correlation is significant at p< 0.01 level (2-tailed).

* Correlation is significant at p< 0.05 level (2-tailed).

Also from the results in Table 3 above, the Mid-range supply chain strategy have a weak but significant relationship (r = 0.591, p<0.05) with firm performance. This is characterized by a nonexistent supply chain strategy in the firm which can lead to total supply chain failure and customer dissatisfaction.

The results of supply chain strategy and firm performance show that it is long-range planning that can support firm competitiveness as evidenced by the effect of long-range supply chain strategies on both supply chain performance and firm performance. The Mid-range strategies are used by most firms as experimental strategies as they craft long-range strategies. The only surprising result is that none of the long-range supply chain is supporting both supply chain performance and firm performance, meaning that firms can implement several supply chain strategies to support various objectives at the supply chain level and corporate level.

3.3. Hypothesis Testing

This study aimed at establishing the relationship between SC strategies and firm performance among large-scale manufacturing firms in Kenya. The literature review and theoretical reasoning led to the reasoning that both Mid-range and Long-range supply chain strategies are associated with firm performance. The four Mid-range supply chain strategies are operational and will affect firm midterm performance. The long-range supply chain strategies are most representative of how companies articulate their models for competing now and in the future. Hence, the following hypotheses were tested:

H: *Supply Chain Strategies are positively related to Firm Performance*

The supply chain strategies items were measured on a scale of 1 to 5 where "5" was to a great extent and "1" to a very small extent. It consisted of statements that sought to measure the extent to which the firms have used the supply chain strategies. Supply chain performance index computed from the achievement on certain items for five years. The Spearman's correlation showed significant relationship between long-range (r = 0. 690, p< 0.01) and mid-range (r = 0.591, p< 0.05) supply chain strategies individually with firm performance. Further analysis using multiple regression analysis is presented in Table 4 below and Annex I.

Table 4: Regression Results for Supply Chain Strategies and Firm Performance

Model Summary: Objective 2 (Data Analysis Model #i) Method: Stepwise (Criteria: Probability-of-F-to-enter≤.050, Probability-of-F-to-remove ≥ .100).					ANOVA(f)		
Stepwise Model	R	R²	Adjusted R²	Std. Error of the Estimate	Mean Square	F	Sig.
1	.545(a)	.297	.291	10.19816	4490.500	43.177	.000(a)
2	.674(b)	.455	.444	9.02904	3432.442	42.104	.000(b)
3	.720(c)	.519	.505	8.52191	2612.157	35.969	.000(c)
4	.757(d)	.574	.556	8.06489	2164.888	33.284	.000(d)
5	.793(e)	.629	.611	7.55540	1900.903	33.300	.000(e)
6	.818(f)	.669	.649	7.17663	1683.812	32.693	.000(f)
7	.837(g)	.701	.679	6.85599	1512.331	32.174	.000(g)
8	.848(h)	.720	.696	6.67527	1358.204	30.481	.000(h)
9	.860(i)	.739	.714	6.47731	1239.438	29.542	.000(i)
10	.868(j)	.754	.727	6.32309	1138.048	28.464	.000(j)
11	.876(k)	.768	.740	6.17524	1053.680	27.631	.000(k)
12	.891(l)	.794	.767	5.84211	999.409	29.282	.000(l)

Source: Research Data, 2014

a, b, c, d, e, f, g, h, i, j, k, l Predictors: (Constant), Supply chain strategies

m Dependent Variable: Firm Performance

From the regression results in Table 4 above, twelve models have been generated using a stepwise approach where the probability-of-F-to-enter was ≤.050 while the probability-of-F-to-remove was ≥ .100. The stepwise multiple regression model number 12 or L is the most significant model since it has the inclusion of most supply chain strategies while the results are significant at the set confidence interval of 95%. Also from the model Table 4, it is observed that as one moves from stepwise model 1 to 12, the standard error of the estimated models decrease from 10.19816 to 5.84211 as so does the F values from 43.177 to 28.282. The adjusted R2 improves from 0.291 to 0.767. Although all the twelve models are significant, stepwise model number twelve is a good predictor of the relationship between supply chain strategies and firm performance.

The stepwise regression model number 12 shows a strong significant relationship between supply chain strategies and firm performance, implying that the supply chain strategies explain 76.7 % of the changes in the firm's performance. The coefficients of this predictive model aimed at addressing the concerns of objective two as modeled in model number eight of the data analysis are given as in Annex I.

From the specific beta coefficients for the measures of supply chain strategies in Annex, I indicate that both long range and Mid-range the supply strategies make some contribution to the firm's performance. All the long range supply chain strategies that affect supply chain performance also affect firm performance. They include the long range risk-hedging supply chain strategy (beta = -0.348); Speed to market supply chain strategy (beta = -0.304) and cash-to-cash cycle supply chain strategy (beta = -0.240). The supply chain strategies that have an impact on firm performance and not the firm's supply chain performance outcome include: efficient supply chain strategies where the firm continuously plans its supply chain network to limit exposure to cost fluctuations(beta = 0.729); project logistics supply chain strategies that allows the firm to cost effectively receives and delivers products as the sources of supply and customer change (beta = 0.435); innovation supply chain strategy focused on variable productivity to meet speculative purchasing and sales promotion (beta = 0.403); Demand supply chain strategy responsive and flexible to customer needs to enable the firm feed customers in ways that are efficient for them (beta = 0.343); Synergistic supply chain strategy where the firm creates additional relationship with supply chain members at the point where their operation interact (beta = 0.261). These are among the long range supply chain strategies that build around projects, innovation, demand forecasting and synergy that are key to the firm's positive performance. The two Mid-range strategies that have a positive impact on the firm's performance include: Third-party supply chain strategy where the firm evaluates opportunities to outsource areas that are not their core competencies in the supply chain (beta = 0.310) and the supply chain strategy where numerous internal and external activities are co-ordinated to conform to the overall business strategy (beta = -0.235). The supply chain strategy on conformance has a negative effect on the firm's performance as it denies the firm some level of flexibility for the firm to adjust to the changes in its environment of operation. It only the investment on long-range supply chain strategies that can lead to improved firm performance levels.

As shown in Table 4 above and Annex I, when Mid-range and long-range supply chain strategies are included in the same model, they have a strong positive effect on firm performance with a correlation coefficient of R = 0.891(l) and adjusted R2 = 0.767, F = 29.282; Sig. = .000(l). This implies that 76.7% of the variance in the firm's performance is explained by the combined variables of Mid-range and long-range supply chain strategies. The relationships between supply chain strategies and firm performance are positive. Given that the calculated F = 29.282, while the F Critical = 1.7611; at α = 5% (95% C.I), numerator degrees of freedom - V1 = 16 (17-1) and denominator degrees of freedom -V2 = 87 (103-16). Then F ≥ F Critical at α = 5%. This is a clear indication that supply chain strategy is a significant predictor of the firm's performance. The relationships explained by the combined variables of Mid-range and long-range supply chain strategies on the firm's performance are positive and statistically significant. Hence, H is accepted.

The second objective had two sub-hypotheses to be tested. Other than the combined effect of supply chain strategies, it was anticipated that Mid-range supply chain strategies are likely to have important implications on firm performance. Hence, the following sub-hypothesis was tested:

H1a: Mid-range SC Strategies are positively related to Firm Performance

The results of Spearman's correlation showed a correlation coefficient of r = 0.591, p< 0.05. The multiple regression analysis is presented in Table 5a&b below.

Table 5a: Regression Model Summary Results for Mid-Range Supply Chain Strategies and Firm Performance

	R	R Square	Std. Error of the Estimate	F	Sig.
Model No. 8a	.545(a)	.297	11.24111	2.218	.102(a)

Source: Research Data, 2014

a Predictors: (Constant), Mid-range Supply chain strategies

b Dependent Variable: Firm Performance

Table 5b: Regression Results for Mid-Range Supply Chain Strategies and Firm Performance

Mid-range Supply chain strategies	Nano-Chain supply chain strategy	Tie down the firm supply chain strategy	Third-party SC strategy	No need for supply chain strategy
Standardized Coefficients (Beta)	.477	.211	-.124	-.175

Source: Research Data, 2014

The beta values for the four Mid-range supply chain strategies show greater individual contributions. From the summary of standardized beta coefficients in Table 5b above, the two Mid-range supply chain strategies that have a positive effect (positive beta value) on the firm's performance outcome include: mid-range supply chain strategies that allows the firm's assets and operations to react to emerging customers trends at each node of the supply chain (Nano-Chain supply chain strategy); and mid-range supply chain strategies where numerous internal and external activities are coordinated to conform to the overall business strategy (Tie down the firm supply chain strategy). The two riskiest Mid-range supply chain strategies for the firm's performance are those mid-range supply chain strategies where the firm evaluates opportunities to outsource areas that are not their core competencies in the supply chain (Third-party SC strategy); and mid-range supply chain strategies where the firm does not have or

pursue a formal supply chain strategy (No need for supply chain strategy).

The analysis in Table 5a and 5b above show that the Mid-range supply chain strategies have a positive relationship with firm performance with a correlation coefficient of R = .545(a) and R2 = 0.297. This implies that 30% of the variance in firm performance is explained by the Mid-range supply chain strategies. Given that α = 5%, the F value of 2.218 is not significant (sign. = 0.102) hence Mid-range supply chain strategies are not good predictors of the firm's performance. Hence, Ha is Accepted.

Hb: Long-range SC Strategies are positively related to Firm Performance

The results of Spearman's correlation showed a correlation coefficient of r = 0. 690, p< 0.01. The multiple regression analysis is presented in Table 6 below.

Table 6: Regression Model Summary Results for Long-Range Supply Chain Strategies and Firm Performance

	R	R Square	Std. Error of the Estimate	F	Sig.
Model No. 8b	.858(a)	.735	8.76469	3.011	.030(a)

Source: Research Data, 2014

a Predictors: (Constant), Long-Range Supply chain strategies

b Dependent Variable: Firm Performance

Based on the beta values for the twelve long-range supply chain strategies that show greater individual contributions to the firm's performance, the strategies that have a positive effect (positive beta value) on firm's performance include: long-range supply chain strategies where the firm continuously plans its supply chain network to limit exposure to cost fluctuations (Efficient Supply Chain strategy, Beta = .610); long-range supply chain strategies focused on variable productivity to meet speculative purchasing and sales promotion (Innovation supply chain strategy, Beta =.421) long-range supply chain strategies responsive and flexible to customer needs to enable the firm Feed Customers in ways that are efficient for them (Demand supply chain strategy, Beta =.368); long-range supply chain strategies that allows the firm to cost effectively receives and delivers products as the sources of supply and customer

change (Project logistics supply chain strategy, Beta =.240); long-range supply chain strategies where the firm creates additional relationship with supply chain members at the point where their operation interact (Synergistic SC strategy, Beta =.183); and lastly long-range supply chain strategies that allows the firm and supply chain members to adopt to different products of different segment of the market (Speed to market supply chain strategy, Beta = -.094).

The most risky long-range supply chain strategies for the firm's performance are those long-range supply chain strategies aimed at speeding and retaining cash flow for the firm (Cash-to-cash cycle supply chain strategy, Beta = -.161); long-range supply chain strategies directed to minimizing risks like production capacity, quality, floods and earthquakes in the process of procurement, production and distribution (Risk-hedging Supply Chain strategy, Beta = -.189); long-range supply chain strategies that provides balance of flexibility and cost efficiency in the supply chain while meeting the requirements of the marketplace requirements (Value chain strategy, Beta = -.211); long-range supply chain strategies that increases the firm's ability to mass-maximize and build close relations with customers when designing new and modifying existing products (Market dominance and backlog supply chain strategy, Beta = -.301); and long-range supply chain strategies that are reactive to procurement, production and distribution in dynamic environments to answer to customer needs (Micro-chain supply chain strategy, Beta = -.404).

The analysis in Table 6 above show that the long-range supply chain strategies have a strong positive effect on firm performance with a correlation coefficient of R = 0. 858 (a) and R2 = 0. 735. This implies that 74% of the variance in firm performance is partly explained by the long-range supply chain strategies. Both Mid-range and long-range supply chain strategies explain 76% of the variance in firm performance, meaning it is the long-range supply chain strategies that contribute most to the firm's performance. Given that $\alpha = 5\%$, the F value of 3.011 is significant (sign. = 0.030) the long-range supply chain strategies are good predictors of the firm's performance. Hence, Hb is Accepted.

4. DISCUSSION OF THE FINDINGS

Scholarly research should contribute to and extend the current literature by filling in existing gaps for both researchers and managers (Kirchoff, 2011; Varadarajan, 2003). This section discusses the findings guided by the primary research objectives and hypotheses. This study aimed at establishing the relationship between SC strategies and performance of large-scale manufacturing firms in Kenya. Two sub-hypotheses were derived from this objective. The hypotheses tested the relation between Mid-range and long-range supply chain strategies with firm performance.

Most previous conceptual and qualitative research has focused on the strategy content with general guidelines, but the influence of these strategies on the relationship between supply chain strategies and firm performance has neither been thoroughly underpinned with theory nor analyzed through empirical research (Chopra & Meindl, 2007). The results

supported the hypothesized relationships except in the case of Mid-range supply chain strategies that don't have much support for firm performance. This empirical evidence is, therefore, a major contribution the specific supply chain strategies (not content) and their relationship firm performance.

Although, most previous studies have examined the concept of supply chain strategy and firm performance, there are very few of them on supply chain strategy and even fewer studies about the relationship between supply chain strategy and firm performance (Gudnason & Riis, 1984). This study has indeed taken this opportunity to confirm that there is a positive and significant relationship between supply chain strategies and firm performance. This clears the contradiction by Menor, Kristal, & Rosenzweig (2007) that the investment in supply chain strategy is associated with increased costs, and it does not translate to improved firm performance.

Mid-range SC strategies are not good predictors of firm performance. Particularly, the supply chain strategies that focus on conformance have a negative effect on the firm's performance as they deny the firm some level of flexibility to adjust to the changes in its environment of operation. It is only the investment on long-range SC strategies that can lead to improved firm performance levels. The riskiest long-range SC strategies for the firm's performance are those long-range supply chain strategies aimed at speeding and retaining cash flow for the firm. This relationship between SC strategies and firm performance is based on data collected over a period of five years. The use of secondary data especially for firm performance metrics is a big strength in explaining the causal relationships. This provided an opportunity to peruse the firm manuals and financial reports to crosscheck the achievements in firm performance. This was to fill the gaps identified by Sánchez & Pérez (2005) which indicated that most studies in firm performance have used cross-sectional data, which are limited in order to explain causal relationships; with failure to use any secondary data to crosscheck firm performance. The study findings agree with the conclusions that supply chain practices have a significant effect on firm performance (Golicic & Smith, 2013; Khan & Pillania, 2008; Lee et al., 2012; Li et al., 2006)

This study used a balanced scorecard to measure firm performance. This affirms that the best way to measure the implementation of any strategy and yield valid results is through the use of the Balanced

Scorecard. This findings are supported by Kaplan & Norton (1992) conclusion that a Balanced Scorecard supports management to improve the financial performance of the enterprise where failure translates the improved operational performance into improved financial performance that send executives back to the drawing board to rethink the company's strategy or its implementation plans.

5. CONCLUSIONS

There is a strong and significant relationship between supply chain strategy and the firm's performance where Supply chain strategies explain 76.7 % of the changes in the firm's performance.

5.1. Contributions to Knowledge

By empirically testing the extent to which supply chain strategies are associated to firm and supply chain performance, the present study adds to academic knowledge in several ways by proving empirical evidence pointing towards the significant use of supply chain strategies that will lead to different levels of achievement in firm performance. The inclusion of the construct supply chain strategy in this study contributes to both the operations management and SCM literature as both the operationalization and empirical testing of supply chain strategy has only been investigated in the strategic management and marketing literature. This study of supply chain strategy within operations management was motivated by premise presented by Boyer & Pagell (2000) and Chan & Qi (2003) that there is need for empirical research that will give an extension of the operations strategy perspective towards the more recent supply chain thinking.

Certain strategies are developed and implemented by firms due to stakeholder pressure, regulatory demands, social legitimacy, and the perceived direct economic benefits. However, the economic benefits and motivation for implementing SC strategies have rarely been tested empirically (Thun & Müller, 2010; Zhu, Sarkis, & Lai, 2008). The findings in this study represent the first empirical study that has found a significant relationship between SC strategies and firm performance improvements. This indicates that firms use SC strategies to improve their financial stewardships, service delivery, operations and customer dynamics that are multiple dimensions related to firm performance. This study also widens the avenue for further research on the moderating role

of SCT on the relationship between SC strategies and firm performance. Scholars can use the results to extend performance metrics, study comparisons of different sample sets, and look at longitudinal data for break-even points on the number of SC strategies and technologies implementation.

The findings from this empirical study provide evidence that the RBV of the firm is an important theory in the study of the relationship between SC strategies to firm performance. This extends the conceptual and empirical research in areas related to SC strategy by suggesting that firms with enough capabilities and resources may be more likely to implement SC strategies and realize improvement in firm performance, compared to the competition. Based on the conclusions by Puri (2013) and that most empirical research on the relationship between supply chain practices and firm performance is limited in number and often with conflicting findings, this current study had set out to conclusively and empirically investigate the role of technology in the relationship between SC strategies and firm performance. This empirical study has contributed to a greater understanding of the relationship between SC strategies and firm performance to the current knowledge in this area.

5.2. Future Research Directions

The limitations in the previous section can be addressed but beyond that, there are a number of interesting and exciting future research possibilities based on the findings from this study. While the objective of this study was achieved, the future research in an effort to enhance the conclusions of this study's findings by focusing on other variables like risk management strategies as a moderator on the relationship between SC strategies and firm performance. This study focused on supply chain strategies that could cut across procurement, value creation and distribution. Future studies can narrow their focus to procurement strategies, value creation strategies, and distribution strategies by comparing their impact on firm performance.

The data collected for firm performance was quantitative in nature. This was in response to Awino's (2011) suggestion that in order to provide a rich research database for future research, future study may explore alternative performance measurement indicators of the quantitative nature, such as financial measures, accounting measures, balance scorecards, linkages to financial statements amongst others. These secondary data was not easy to get. The firms indicated that it was classified information while other indicated that was confidential, hence giving the researcher tough conditions in its use.

6. REFERENCES

Agha, S., Alrubaiee, L., & Jamhour, M. (2011). Effect of Core Competence on Competitive Advantage and Organizational Performance. International Journal of Business and Management, 7(1), 192–204. http://doi.org/10.5539/ijbm.v7n1p192

Bhagwat, R., & Sharma, M. K. (2007). Performance measurement of supply chain management: A balanced scorecard approach. Computers and Industrial Engineering, 53(1), 43–62. http://doi.org/10.1016/j.cie.2007.04.001

Bolo, A. Z. (2011). An empirical investigation of selected strategy variables on firms performance : A study of supply chain management in large private manufacturing firms in Kenya. Business Administration and Management, 3(October), 228–236.

Boyer, K. K., & Pagell, M. (2000). Measurement issues in empirical research: Improving measures of operations strategy and advanced manufacturing technology. Journal of Operations Management, 18(3), 361–374. http://doi.org/10.1016/S0272-6963(99)00029-7

Burgess, K., Singh, P. J., & Koroglu, R. (2006). Supply chain management: a structured literature review and implications for future research. International Journal of Operations & Production Management, 26(7), 703–729. http://doi.org/10.1108/01443570610672202

Chan, F. T. S., & Qi, H. J. (2003). An innovative performance measurement method for supply chain management. Supply Chain Management: An International Journal, 8(3), 209–223. http://doi.org/10.1108/13598540310484618

Chopra, S., & Meindl, P. (2007). Supply Chain Management: Strategy, Planning and Operation, (3rd ed.). Pearson Prentice Hall.

Collins, J. D., Worthington, W. J., Reyes, P. M., & Romero, M. (2010). Knowledge management, supply chain technologies, and firm performance. Management Research Review, 33(10), 947–960. http://doi.org/10.1108/01409171011083969

Cooper, M. C., & Ellram, L. M. (1993). Characteristics of Supply Chain Management and the Implications for Purchasing and Logistics Strategy. The International Journal of Logistics Management, 4(2), 13–24. http://doi.org/10.1108/09574099310804957

Cousins, P. D., Lawson, B., & Squire, B. (2006). Supply chain management: theory and practice – the emergence of an academic discipline? International Journal of Operations & Production Management, 26(7), 697–702. http://doi.org/10.1108/01443570610672194

Ezirim, A. C. ., & Nwokah, N. G. . (2009). Firms entrepreneurial orientation and export marketing performance in the Nigerian non-oil sector. European Journal of Scientific Research, 35(3), 318–336.

Fawcett, S. E., Ogden, J. a., Magnan, G. M., & Cooper, M. B. (2006). Organizational commitment and governance for supply chain success. International Journal of Physical Distribution & Logistics Management, 36(1), 22–35. http://doi.org/10.1108/09600030610642913

Gattorna, J. (2006). Living supply chains: how to mobilize the enterprise around delivering what your customers want. Financial Times Series (1st ed.). Prentice Hall. http://doi.org/0-273-70614-4

GoK. (2007). Kenya Vision 2030: A Globally Competitive and Prosperous Kenya. Government of Kenya, National Economic and Social Council (NESC). Nairobi, Kenya.

Golicic, S. L., & Smith, C. D. (2013). A Meta-Analysis of Environmentally Sustainable Supply Chain Management Practices and Firm Performance. Journal of Supply Chain Management, 49(2), 78–95.

Gudnason, C., & Riis, J. (1984). Manufacturing strategy. Omega (Vol. 12). http://doi.org/10.1016/0305-0483(84)90057-4

Gunasekaran, A., Patel, C., & Tirtiroglu, E. (2001). Performance measures and metrics in a supply chain environment. International Journal of Operations & Production Management, 21(1/2), 71–87. http://doi.org/10.1108/01443570110358468

Halley, A., & Beaulieu, M. (2009). Mastery of operational competencies in the context of supply chain management. Supply Chain Management: An International Journal, 14(1), 49–63. http://doi.org/10.1108/13598540910927304

Hines, T. (2009). Supply chain strategies: customer driven and customer focussed (2nd ed.). Routledge. http://doi.org/10.4324/9780203631669

Hult, G. T. M., Ketchen, D. J., Cavusgil, S. T., & Calantone, R. J. (2006). Knowledge as a strategic resource in supply chains. Journal of Operations Management, 24(5), 458–475. http://doi.org/10.1016/j.jom.2005.11.009

Johnson, G., Scholes, K., & Whittington, R. (2007). Exploring Corporate Strategy. (8th ed.). Prentice Hall. http://doi.org/10.1016/0142-694X(85)90029-8

Kamaruddin, N. K., & Udin, Z. M. (2009). Supply chain technology adoption in Malaysian automotive suppliers. Journal of Manufacturing Technology Management, 20(3), 385–403. http://doi.org/10.1108/17410380910936819

Kaplan, R. S., & Norton, D. P. (1992). The balanced scorecard - measure that drive performance. Harvard Business Review, 70(1), 71–79.

Kate Cowles. (2006). Statistical Methods and Computing: Sample size for confidence intervals with known t Intervals. IOWA.

Keegan, D. P., Eiler, R. G., & Charles, R. J. (1989). Are your performance measures obsolete? Management Accounting, 70(12), 45–50. http://doi.org/10.1177/004057368303900411

Ketchen, D. J., & Giunipero, L. C. (2004). The intersection of strategic management and supply chain management. Industrial Marketing Management, 33(1), 51–56. http://doi.org/10.1016/j.indmarman.2003.08.010

Khan, A., & Pillania, R. (2008). Strategic sourcing for supply chain agility and firms' performance: A study of Indian manufacturing sector. Management Decision, 46(10), 1508–1530. http://doi.org/10.1108/00251740810920010

Kidombo, H. (2007). Human Resource Strategic Orientation, Organizational Commitment and Firm Performance in Large Private Manufacturing Firms in Kenya. University of Nairobi.

Kirchoff, J. F. (2011). A Resource-Based Perspective on Green Supply Chain Management and Firm Performance. University of Tennessee.

Lars-Erik Gadde, Hakan Hakansson, G. P. (2001). Supply Network Strategies (2nd ed.). West Sussex, UK: John Wiley & Sons.

Lee, S. M., Kim, S. T., & Choi, D. (2012). Green supply chain management and organizational performance. Industrial Management & Data Systems, 112(8), 1148–1180. http://doi.org/DOI 10.1108/02635571211264609

Li, S., Ragu-Nathan, B., Ragu-Nathan, T. S., & Subba Rao, S. (2006). The impact of supply chain management practices on competitive advantage and organizational performance. Omega, 34(2), 107–124. http://doi.org/10.1016/j.omega.2004.08.002

Menor, L. J., Kristal, M. M., & Rosenzweig, E. D. (2007). Examining the Influence of Operational Intellectual Capital on Capabilities and Performance. Manufacturing & Service Operations Management, 9(4), 559–578. http://doi.org/10.1287/msom.1060.0131

Mentzer, J. T., DeWitt, W., Keebler, J. S., Min, S., Nix, N. W., Smith, C. D., & Zacharia, Z. G. (2001). Defining Supply Chain Management. Journal of Business Logistics, 22(2), 1–25. http://doi.org/10.1002/j.2158-1592.2001.tb00001.x

Neely, A., Bourne, M., & Kennerley, M. (2000). Performance measurement system design: developing and testing a process-based approach. International Journal of Operations & Production Management, 20(10), 1119–1145. http://doi.org/10.1108/01443570010343708

Neuman, W. L. (2009). Social Research Methods: Qualitative and Quantitative Approaches (7th ed.). Pearson.

Okoth, J. (2012, February 17). Are multinationals leaving Kenya? The Financial Post: Bruce House, Standard Street. Nairobi.

Ou, C. S., Liu, F. C., Hung, Y. C., & Yen, D. C. (2010). A structural model of supply chain management on firm performance. International Journal of Operations & Production Management, 30(5), 526–545. http://doi.org/10.1108/01443571011039614

Patterson, K. a., Grimm, C. M., & Corsi, T. M. (2003). Adopting new technologies for supply chain management. Transportation Research, E 39, 95–121. http://doi.org/10.1016/S1366-5545(02)00041-8

Puri, G. (2013). Factors affecting the Adoption of B2B E-commerce – An Empirical Study, 6(1), 13–22.

PwCIL. (2010). Industrial Manufacturing: A brief overview of the industrial manufacturing sector in Kenya. Nairobi, Kenya.

Qrunfleh, S., & Tarafdar, M. (2014). Supply chain information systems strategy: Impacts on supply chain performance and firm performance. International Journal of Production Economics, 147(PART B), 340–350. http://doi.org/10.1016/j.ijpe.2012.09.018

Russell, D. M., & Hoag, A. M. (2004). People and information technology in the supply chain: Social and organizational influences on adoption. International Journal of Physical Distribution & Logistics Management, 34(2), 102–122. http://doi.org/10.1108/09600030410526914

Sánchez, A. M., & Pérez, M. P. (2005). Supply chain flexibility and firm performance: A conceptual model and empirical study in the automotive industry. International Journal of Operations & Production Management, 25(7), 681–700. http://doi.org/10.1108/01443570510605090

Stiles, J. (2003). A philosophical justification for a realist approach to strategic alliance research. Qualitative Market Research: An International Journal, 6(4), 263–271. http://doi.org/10.1108/13522750310495346

Teeratansirikool, L., Siengthai, S., Badir, Y., & Charoenngam, C. (2013). Competitive strategies and firm performance: the mediating role of performance measurement. International Journal of Productivity and Performance Management, 62(2), 168–184. http://doi.org/10.1108/17410401311295722

Thun, J.-H., & Müller, A. (2010). An Empirical Analysis of Green Supply Chain Management in the German Automotive Industry. Business Strategy and the Environment, 19(2), 119–132. http://doi.org/10.1002/bse

Tomaskovic-devey, D., Leiter, J., Thompson, S., Quarterly, A. S., Sep, N., & Tomaskovic-, D. (2007). Organizational Survey Nonresponse Organizational Survey Nonresponse Devey , Shealy Thompson. Science, 39(3), 439–457.

Umma, S. (2006). Research methods for business (4th ed.). Wiley India Pvt.

Varadarajan, P. R. (2003). Musings on Relevance and Rigor of scholarly research in marketing. Journal of the Academy of Marketing Science, 31(4), 368–376.

Varma, S., Wadhwa, S., & Deshmukh, S. G. (2006). Implementing supply chain management in a firm: issues and remedies. Asia Pacific Journal of Marketing and Logistics, 18(3), 223–243. http://doi.org/10.1108/13555850610675670

Wilson, E. J., & Lilien, G. L. (1992). Using single informants to study group choice: An examination of research practice in organizational buying. Marketing Letters, 3(3), 297–305. http://doi.org/10.1007/BF00994137

Zhu, Q., Sarkis, J., & Lai, K. (2008). Green supply chain management implications for "closing the loop." Transportation Research Part E: Logistics and Transportation Review, 44(1), 1–18. http://doi.org/10.1016/j.tre.2006.06.003

Zolait, A. H., Ibrahim, A. R., Chandran, V. G. R., Pandiyan, V., & Sundram, K. (2010). Supply chain integration: an empirical study on manufacturing industry in Malaysia. Journal of Systems and Information Technology, 12(3), 210–221. http://doi.org/10.4018/jtd.2010070103

Annex I: Regression Coefficients (a) for Supply Chain Strategies and Firm Performance

Model	Indicators: Objective i (Data Analysis Model #i)	Unstandardized Coefficients		Standardized Coefficients	t	Sig.
		B	Std. Error	Beta		
Model # 1						
	(Constant)	29.97	4.746		6.315	.000
A SC strategy focused on variable productivity to meet speculative purchasing and sales promotion: LR Innovation SC strategy	7.70	1.171	.545	6.571	.000	
Model # 2	(Constant)	**-58.61**	**16.941**		**-3.459**	**.001**
A SC strategy focused on variable productivity to meet speculative purchasing and sales promotion: LR Innovation SC strategy	7.77	1.037	.551	7.491	.000	
SC a strategy responsive and flexible to customer needs to enable the firm Feed Customers in ways that are efficient for them: LR Demand SC strategy.	17.93	3.323	.397	5.397	.000	
Model # 3	(Constant)	**-24.41**	**18.522**		**-1.318**	**.191**
A SC strategy focused on variable productivity to meet speculative purchasing and sales promotion: LR Innovation SC strategy	9.58	1.096	.679	8.734	.000	
A SC a strategy responsive and flexible to customer needs to enable the firm Feed Customers in ways that are efficient for them: LR Demand SC strategy.	16.52	3.160	.365	5.227	.000	
A SC strategy that is reactive to procurement, production and distribution in dynamic environments to answer to customer needs: LR Micro-chain SC strategy	-7.15	1.956	-.286	-3.658	.000	
Model # 4	(Constant)	**-32.42**	**17.673**		**-1.835**	**.070**
A SC strategy focused on variable productivity to meet speculative purchasing and sales promotion: LR Innovation supply chain strategy	7.01	1.264	.497	5.543	.000	
SC a strategy responsive and flexible to customer needs to enable the firm Feed Customers in ways that are efficient for them: LR Demand SC strategy.	17.44	3.002	.386	5.809	.000	
A SC strategy that is reactive to procurement, production and distribution in dynamic environments to answer to customer needs: LR Micro-chain SC strategy	-8.24	1.876	-.330	-4.395	.000	
A strategy that allows the firm to cost effectively receives and delivers products as the sources of supply and customer change: LR Project logistics SC strategy	5.02	1.410	.312	3.557	.001	
Model # 5	(Constant)	**11.64**	**20.132**		**.578**	**.564**
A supply chain strategy focused on variable productivity to meet speculative purchasing and sales promotion: LR Innovation SC strategy	7.19	1.185	.510	6.070	.000	
SC strategy responsive and flexible to customer needs to enable the firm Feed Customers in ways	15.69	2.849	.347	5.507	.000	

Model	Indicators: Objective i (Data Analysis Model #i)	Unstandardized Coefficients		Standardized Coefficients	t	Sig.
		B	Std. Error	Beta		
that are efficient for them: LR Demand SC strategy.						
A SC strategy that is reactive to procurement, production and distribution in dynamic environments to answer to customer needs: LR Micro-chain SC strategy	-9.64	1.795	-.386	-5.373	.000	
A strategy that allows the firm to cost effectively receives and delivers products as the sources of supply and customer change: LR Project logistics SC strategy	6.51	1.377	.405	4.727	.000	
A SC strategy aimed at speeding and retaining cash flow for the firm: LR Cash-to-cash cycle SC strategy	-7.36	1.913	-.257	-3.847	.000	
Model # 6	**(Constant)**	**22.55**	**19.389**		**1.163**	**.248**
A SC strategy focused on variable productivity to meet speculative purchasing and sales promotion: LR Innovation SC strategy	7.18	1.126	.509	6.374	.000	
SC a strategy responsive and flexible to customer needs to enable the firm Feed Customers in ways that are efficient for them: LR Demand SC strategy.	14.34	2.734	.317	5.245	.000	
A SC strategy that is reactive to procurement, production and distribution in dynamic environments to answer to customer needs: LR Micro-chain SC strategy	-8.45	1.740	-.338	-4.853	.000	
A strategy that allows the firm to cost effectively receives and delivers products as the sources of supply and customer change: LR Project logistics SC strategy	7.81	1.362	.486	5.732	.000	
A SC strategy aimed at speeding and retaining cash flow for the firm: LR Cash-to-cash cycle SC strategy	-7.13	1.818	-.249	-3.921	.000	
A strategy that increases the firm's ability to mass-maximize and build close relations with customers when designing new and modifying existing products: LR Market dominance and backlog SC strategy	-3.95	1.159	-.232	-3.408	.001	
Model # 7	**(Constant)**	**28.09**	**18.603**		**1.510**	**.134**
A supply chain strategy focused on variable productivity to meet speculative purchasing and sales promotion: LR Innovation SC strategy	6.99	1.077	.496	6.497	.000	
Supply chain a strategy responsive and flexible to customer needs to enable the firm Feed Customers in ways that are efficient for them: LR Demand SC strategy.	12.752	2.659	.282	4.796	.000	
A SC strategy that is reactive to procurement, production and distribution in dynamic environments to answer to customer needs: LR	-9.31	1.684	-.373	-5.529	.000	

Model	Indicators: Objective i (Data Analysis Model #i)	Unstandardized Coefficients		Standardized Coefficients	t	Sig.
		B	Std. Error	Beta		
Micro-chain SC strategy						
A strategy that allows the firm to cost effectively receives and delivers products as the sources of supply and customer change: LR Project logistics SC strategy	7.42	1.307	.462	5.679	.000	
A SC strategy aimed at speeding and retaining cash flow for the firm: LR Cash-to-cash cycle SC strategy	-7.276	1.737	-.254	-4.188	.000	
A strategy that increases the firm's ability to mass-maximize and build close relations with customers when designing new and modifying existing products: LR Market dominance and backlog SC strategy	-4.63	1.127	-.271	-4.103	.000	
A strategy where the firm evaluates opportunities to outsource areas that are not their core competencies in the supply chain: MR Third-party SC strategy	3.19	.997	.199	3.207	.002	
Model # 8	**(Constant)**	**26.26**	**18.127**		**1.449**	**.151**
A supply chain strategy focused on variable productivity to meet speculative purchasing and sales promotion: LR Innovation SC strategy	6.81	1.051	.483	6.480	.000	
SC a strategy responsive and flexible to customer needs to enable the firm Feed Customers in ways that are efficient for them: LR Demand SC strategy.	13.39	2.601	.296	5.147	.000	
A SC strategy that is reactive to procurement, production and distribution in dynamic environments to answer to customer needs: LR Micro-chain SC strategy	-9.64	1.645	-.386	-5.861	.000	
A strategy that allows the firm to cost effectively receives and delivers products as the sources of supply and customer change: LR Project logistics supply chain strategy	6.09	1.378	.379	4.420	.000	
A SC strategy aimed at speeding and retaining cash flow for the firm: LR Cash-to-cash cycle SC strategy	-7.06	1.694	-.247	-4.170	.000	
A strategy that increases the firm's ability to mass-maximize and build close relations with customers when designing new and modifying existing products: LR Market dominance and backlog SC strategy	-5.73	1.183	-.336	-4.845	.000	
A strategy where the firm evaluates opportunities to outsource areas that are not their core competencies in the supply chain: MR Third-party SC strategy	3.012	.973	.188	3.102	.003	
A strategy where the firm continuously plans its supply chain network to limit exposure to cost fluctuations: LR Efficient SC strategy	2.56	1.023	.200	2.504	.014	
Model # 9	**(Constant)**	**26.41**	**17.590**		**1.502**	**.137**
A SC strategy focused on variable productivity to meet speculative purchasing and sales	7.19	1.030	.510	6.982	.000	

Model	Indicators: Objective i (Data Analysis Model #i)	Unstandardized Coefficients		Standardized Coefficients	t	Sig.
		B	Std. Error	Beta		
promotion: LR Innovation SC strategy						
SC a strategy responsive and flexible to customer needs to enable the firm Feed Customers in ways that are efficient for them: LR Demand SC strategy.	14.26	2.546	.315	5.600	.000	
A supply chain strategy that is reactive to procurement, production and distribution in dynamic environments to answer to customer needs: LR Micro-chain supply chain strategy	-9.89	1.599	-.396	-6.185	.000	
A strategy that allows the firm to cost effectively receives and delivers products as the sources of supply and customer change: LR Project logistics SC strategy	7.76	1.480	.483	5.241	.000	
A SC strategy aimed at speeding and retaining cash flow for the firm: LR Cash-to-cash cycle SC strategy	-8.65	1.752	-.303	-4.940	.000	
A strategy that increases the firm's ability to mass-maximize and build close relations with customers when designing new and modifying existing products: LR Market dominance and backlog SC strategy	-6.50	1.185	-.381	-5.486	.000	
A strategy where the firm evaluates opportunities to outsource areas that are not their core competencies in the SC: MR Third-party SC strategy	4.51	1.103	.281	4.094	.000	
A strategy where the firm continuously plans its SC network to limit exposure to cost fluctuations: LR Efficient SC strategy	3.83	1.104	.299	3.471	.001	
A SC strategy that allows the firm and supply chain members to adopt to different products of different segment of the market: LR Speed to market SC strategy.	-2.89	1.101	-.258	-2.626	.010	
Model # 10	**(Constant)**	**38.23**	**17.878**		**2.139**	**.035**
A SC strategy focused on variable productivity to meet speculative purchasing and sales promotion: LR Innovation SC strategy	6.45	1.054	.457	6.116	.000	
SC strategy responsive and flexible to customer needs to enable the firm Feed Customers in ways that are efficient for them: LR Demand SC strategy.	13.97	2.488	.309	5.616	.000	
A SC strategy that is reactive to procurement, production and distribution in dynamic environments to answer to customer needs: LR Micro-chain SC strategy	-10.32	1.572	-.413	-6.567	.000	
A strategy that allows the firm to cost effectively receives and delivers products as the sources of supply and customer change: LR Project logistics SC strategy	8.56	1.484	.533	5.769	.000	

| Model | Indicators: Objective i (Data Analysis Model #i) | Unstandardized Coefficients | | Standardized Coefficients | t | Sig. |
		B	Std. Error	Beta		
Third-party SC strategy						
A strategy where the firm continuously plans its SC network to limit exposure to cost fluctuations: LR Efficient SC strategy	4.82	1.095	.376	4.400	.000	
A SC strategy that allows the firm and SC members to adopt to different products of different segment of the market: LR Speed to market SC strategy.	-3.39	1.061	-.303	-3.195	.002	
A SC strategy where numerous internal and external activities are co-ordinated to conform to the overall business strategy: MR Tie down the firm SC strategy	-3.64	1.351	-.162	-2.690	.008	
A strategy where the firm creates additional relationship with SC members at the point where their operation interact: LR Synergistic SC strategy	2.22	.945	.161	2.347	.021	
Model # 12	**(Constant)**	**55.14**	**17.918**		**3.078**	**.003**
A SC strategy focused on variable productivity to meet speculative purchasing and sales promotion: LR Innovation SC strategy	5.69	1.009	.403	5.633	.000	
SC strategy responsive and flexible to customer needs to enable the firm Feed Customers in ways that are efficient for them: LR Demand SC strategy.	15.53	2.367	.343	6.561	.000	
A SC strategy that is reactive to procurement, production and distribution in dynamic environments to answer to customer needs: LR Micro-chain SC strategy	-13.02	1.597	-.521	-8.155	.000	
A strategy that allows the firm to cost effectively receives and delivers products as the sources of supply and customer change: LR Project logistics SC strategy	6.99	1.452	.435	4.817	.000	
A SC strategy aimed at speeding and retaining cash flow for the firm: LR Cash-to-cash cycle SC strategy	-6.87	1.669	-.240	-4.119	.000	
A strategy that increases the firm's ability to mass-maximize and build close relations with customers when designing new and modifying existing products: LR Market dominance and backlog SC strategy	-7.321	1.179	-.429	-6.207	.000	
A strategy where the firm evaluates opportunities to outsource areas that are not their core competencies in the supply chain: MR Third-party SC strategy	4.989	1.007	.310	4.951	.000	
A strategy where the firm continuously plans its supply chain network to limit exposure to cost fluctuations: LR Efficient SC strategy	9.35	1.678	.729	5.572	.000	
A SC strategy that allows the firm and supply	-3.41	1.004	-.304	-3.393	.001	

Model	Indicators: Objective i (Data Analysis Model #i)	Unstandardized Coefficients		Standardized Coefficients		
		B	Std. Error	Beta	t	Sig.
chain members to adopt to different products of different segment of the market: LR Speed to market SC strategy.						
A SC strategy where numerous internal and external activities are co-ordinated to conform to the overall business strategy: MR Tie down the firm SC strategy	-5.26	1.363	-.235	-3.857	.000	
A strategy where the firm creates additional relationship with SC members at the point where their operation interact: LR Synergistic SC strategy	3.601	.980	.261	3.673	.000	
A SC strategy directed to minimizing risks like production capacity, quality, floods and earthquakes in the process of procurement, production and distribution: LR Risk-hedging SC strategy	-6.42	1.868	-.348	-3.434	.001	

Source: Research Data, 2014

a Dependent Variable: Firm Performance

Method: Stepwise (Criteria: Probability-of-F-to-enter≤.050, Probability-of-F-to-remove≥ .100).

Auhtor's Biography:

Richard Nyaoga is a Lecturer in the Faculty of Commerce Department of Accounting, Finance and Management Science of Egerton University-Kenya. Richard has a wide experience in teaching spanning over seven years. Richard has widely published in various peer reviewed journals. Richard has an Undergraduate and Masters Degrees from The University of Nairobi Kenya and a PhD in Management Science.

Peterson Magutu is a Lecturer in The University of Nairobi. He has taught, published and consulted widely on Management Science related areas. Dr. Magutu has An Undergraduate, Master and PhD both from the University of Nairobi. Magutu's research interests lie in developing and testing basic models used in operations management and management science especially in supply chain management.

Josiah Aduda is currently the Associate Professor and Dean of the school of business University of Nairobi. Dr. Aduda has wide experience in teaching and publishing in areas of Management. He has undergraduate & Masters from University of Nairobi and a PhD from Univeristy of Dar es Salaam- Tanzania.

Permissions

The contributors of this book come from diverse backgrounds, making this book a truly international effort. This book will bring forth new frontiers with its revolutionizing research information and detailed analysis of the nascent developments around the world.

We would like to thank all the contributing authors for lending their expertise to make the book truly unique. They have played a crucial role in the development of this book. Without their invaluable contributions this book wouldn't have been possible. They have made vital efforts to compile up to date information on the varied aspects of this subject to make this book a valuable addition to the collection of many professionals and students.

This book was conceptualized with the vision of imparting up-to-date information and advanced data in this field. To ensure the same, a matchless editorial board was set up. Every individual on the board went through rigorous rounds of assessment to prove their worth. After which they invested a large part of their time researching and compiling the most relevant data for our readers.

The editorial board has been involved in producing this book since its inception. They have spent rigorous hours researching and exploring the diverse topics which have resulted in the successful publishing of this book. They have passed on their knowledge of decades through this book. To expedite this challenging task, the publisher supported the team at every step. A small team of assistant editors was also appointed to further simplify the editing procedure and attain best results for the readers.

Apart from the editorial board, the designing team has also invested a significant amount of their time in understanding the subject and creating the most relevant covers. They scrutinized every image to scout for the most suitable representation of the subject and create an appropriate cover for the book.

The publishing team has been an ardent support to the editorial, designing and production team. Their endless efforts to recruit the best for this project, has resulted in the accomplishment of this book. They are a veteran in the field of academics and their pool of knowledge is as vast as their experience in printing. Their expertise and guidance has proved useful at every step. Their uncompromising quality standards have made this book an exceptional effort. Their encouragement from time to time has been an inspiration for everyone.

The publisher and the editorial board hope that this book will prove to be a valuable piece of knowledge for researchers, students, practitioners and scholars across the globe.

List of Contributors

Douglas Steven Hill
University of Southampton

Juan Francisco Zurita Duque and HelleSkøtt
Novozymes A/S

Moema Pereira Nunes, Luciana Marques Vieiraand José Antônio Valle Antunes Jr.
Unisinos

ArunaApteand Keenan D. Yoho
Naval Postgraduate School

Cullen M. Greenfield and Cameron A. Ingram
United States Navy

Gustavo CorrêaMirapalheta and FlaviaJunqueira de Freitas
FundaçãoGetulio Vargas – EAESP

Luis Cesar Mondini
University Center Leonardo Da Vinci

Denise Del PráNetto Machado
Universidade Regional de Blumenau

Marcia Regina Santiago Scarpin
FundaçãoGetúlio Vargas

AlaaAljunaidi
Pepsi Cola - Bugshan (SIBCO)

Samuel Ankrah
Specialist Health Ghana

João Victor Rodrigues Silva
Center for Graduate Studies and Research in Business Administration (CEPEAD)/Federal University of Minas Gerais (UFMG)

Ricardo Silveira Martins
UFMG

DivyaChoudharyand JitendraMadaan
Indian Institute of Technology

RakeshNarain
Motilal Nehru National Institute of Technology

PriscilaLaczynski de Souza Miguel,Eliane Pereira ZamithBrito, LilianSoares Pereira Carvalho and Manuel de Andrade e Silva Reis
FGV-E A E SP

Mats Deleryd, Anders Fundin and Anna Sannö
Department of Product Realisation, Mälardalen University, Sweden

Richard BitangeNyaoga
Egerton University

Peterson Obara Magutuand Josiah Aduda
University of Nairobi

Index